Concentration and Power in the Food System

Contemporary Food Studies: Economy, Culture and Politics

Series Editors: David Goodman and Michael K. Goodman
ISSN: 2058–1807

This interdisciplinary series represents a significant step toward unifying the study, teaching, and research of food studies across the social sciences. The series features authoritative appraisals of core themes, debates and emerging research, written by leading scholars in the field. Each title offers a jargon-free introduction to upper-level undergraduate and postgraduate students in the social sciences and humanities.

Further titles forthcoming

Concentration and Power in the Food System

Who Controls What We Eat?

Philip H. Howard

Bloomsbury Academic
An imprint of Bloomsbury Publishing Plc

B L O O M S B U R Y
LONDON • OXFORD • NEW YORK • NEW DELHI • SYDNEY

Bloomsbury Academic
An imprint of Bloomsbury Publishing Plc

50 Bedford Square	1385 Broadway
London	New York
WC1B 3DP	NY 10018
UK	USA

www.bloomsbury.com

BLOOMSBURY and the Diana logo are trademarks of Bloomsbury Publishing Plc

First published 2016
Reprinted 2016 (twice)

British Library Cataloguing-in-Publication Data
A catalogue record for this book is available from the British Library.

ISBN: HB: 978-1-4725-8112-9
       PB: 978-1-4725-8111-2
       ePDF: 978-1-4725-8113-6
       ePub: 978-1-4725-8114-3

Library of Congress Cataloging-in-Publication Data
Howard, Philip H., 1971- author.
Concentration and power in the food system : who controls what we eat? /
by Philip H. Howard.
pages cm. – (Contemporary food studies: economy, culture and
politics, ISSN 2058-1807 ; volume 3)
ISBN 978-1-4725-8111-2 (pbk.) – ISBN 978-1-4725-8112-9 (hardback)
1. Food industry and trade–United States. 2. Food supply–United States. I. Title. II. Series:
Contemporary food studies: economy, culture and politics ; v. 3. 2058-1807
HD9005.H68 2016
338.10973–dc23
2015025464

Series: Contemporary Food Studies: Economy, Culture and Politics, 2058–1807, volume 3

Typeset by Integra Software Services Pvt. Ltd.
Printed and bound in the United States of America

Contents

List of Figures, Tables, and Boxes

Figures

Tables

Acknowledgments

I am incredibly fortunate to have worked at three universities with numerous food and agriculture scholars: the University of Missouri, University of California at Santa Cruz, and Michigan State University. Each of these experiences played an important role in development of ideas in this book.

Missouri had a formative influence, particularly my experience studying food industry concentration with Bill Heffernan and Mary Hendrickson. My views were also shaped by other members of the Sociology of Agriculture Working Group, including Beth Barham, Jere Gilles, John Green, Bob Gronski, Judy Heffernan, John Ikerd, David Lind, Anna Kleiner, Bryce Oates, Andy Raedeke, and Sandy Rikoon, In addition, I benefited from interactions with former members, especially Doug Constance and Leland Glenna.

Participating in the Agro-Food Studies Working Group in Santa Cruz was an unparalleled experience, providing me the opportunity to participate in the exchange of ideas between Patricia Allen, Chris Bacon, Melanie DuPuis, Margaret Fitzsimmons, Bill Friedland, Brian Fulfrost, Tara Pisani Gareau, David Goodman, Mike Goodman, Julie Guthman, Jill Harrison, Hilary Melcarek, Albie Miles, Katie Monsen, Dustin Mulvaney, Jan Perez, Diana Stuart, and Keith Warner.

At Michigan State University I continued to learn much from colleagues, including Jim Bingen, Larry Busch, Laura DeLind, Mike Hamm, Craig Harris, Dan Jaffee, Paul Thompson, Kyle Whyte, David Wright, and Wynne Wright. Michigan State University also granted me a sabbatical leave in 2013–2014 to complete this manuscript.

I wish to thank University of Utah, Division of Nutrition for hosting me during this sabbatical, as well as providing opportunities to refine ideas through seminars and guest lectures. Special thanks go to Sally and Howard Ogilvie for important contributions to this project and Ginger Ogilvie for her tremendous support and patience. Finally, I thank series editors David Goodman and Mike Goodman for pointing me to many relevant resources and providing extremely helpful suggestions for improvement.

Chapter 1

Food system concentration: a political economy perspective

> *Power is not a means, it's an end.*
>
> —O'Brien (*Nineteen Eighty-Four*)

If you go into a typical grocery store in the United States and make your way back to the margarine case, you will probably see approximately a dozen different brands. If you look very closely at the packaging, however, you may find a small seal, which signifies the majority of these are owned by either Unilever or ConAgra (Table 1.1). Although these two firms dominate the margarine market—Unilever accounts for 51.2 percent of sales in the US market and ConAgra for 16.9 percent (Grocery Headquarters 2013)—their power is hidden from us through an illusion of numerous competing brands. Margarine is not a unique case, and while the number of options offered may differ, similar patterns can be found in almost every food or beverage category. The bread shelves, for example, may provide slightly more choices, but this conceals the fact that Grupo Bimbo and Flowers Foods each own more than a dozen leading brands and together control approximately half of the US market (Thomas and Cavale 2013). The wine aisle may contain literally hundreds of brands, but it is very

Table 1.1 Ownership of margarine brands

Unilever	ConAgra
Becel	Blue Bonnet
Brummel & Brown	Fleischmann's
I Can't Believe It's Not Butter	Move Over Butter
Imperial	Parkay
Promise	
Shedd's Spread Country Crock	

difficult to discern that scores of these, as well as more than half of US sales, are controlled by only three companies: Gallo, The Wine Group, and Constellation (Howard et al. 2012). In nearly every other stage of the food system, including retailing, distribution, farming and farm inputs (e.g., seeds, fertilizers, pesticides), a limited number of firms or operations tend to make up the vast majority of sales.

Is this a problem? An increasing number of people argue that indeed it is: the firms that dominate these industries are criticized for a long list of purported negative impacts on society and the environment. Just a few examples include:

- Walmart, which controls 33 percent of US grocery retailing, is challenged for exploiting its suppliers, taking advantage of taxpayer subsidies, and paying extremely low worker wages.

- McDonald's, which controls more than 18 percent of US fast food sales, is also critiqued for extremely low wages, as well as the negative health consequences and environmental impacts of its products.

- Tyson, which controls more than 17 percent of US chicken, pork, and beef processing is reproached for its pollution, poor treatment of farmers, and contributions to the decline of rural communities.

- Monsanto, which controls 26 percent of the global commercial seed market, is denounced for its influence on government policies, spying on farmers it suspects of saving and replanting seeds, and the environmental impacts of herbicides tied to these seeds.

These impacts tend to disproportionately affect the disadvantaged—such as women, young children, recent immigrants, members of minority ethnic groups, and those of lower socioeconomic status—and as a result, reinforce existing inequalities (Allen and Wilson 2008). Like ownership relations, the full extent of these consequences may be hidden from public view.

This book seeks to illuminate which firms have become the most dominant, and more importantly, how they shape and reshape society in their efforts to increase their control. These dynamics have received insufficient attention from academics and even critics of the current food system. The power of dominant firms extends far beyond narrow economic boundaries, for example, providing them with the ability to damage numerous communities and ecosystems in their pursuit of higher than average profits. The social resistance provoked by these negative consequences is another area that is less visible to the majority of the population. When such resistance is evident at all, it frequently appears insignificant, failing to challenge the direction of current trends. Even very small movements, however, may influence which firms end up winners or losers or close off particular avenues for growth. These accomplishments also suggest potential limits and therefore the possibility that dominant firms may experience much greater threats to their power in the future.

Increasing concentration

Concentration is a term used to describe the composition of a given market, and especially its potential impacts on competition. At one end of the spectrum are markets that are described as unconcentrated or fragmented, which economists consider to be freely competitive (Figure 1.1). In this type of market, sellers are "price takers" and lack the ability to raise prices. At the other end of the spectrum are concentrated markets, which in their most extreme form are monopolies controlled by just one firm. In these situations there are no alternatives, and the monopolists have substantial power to raise prices without losing customers. Also at this end of the spectrum are oligopolies, in which markets are dominated by several large firms but are characterized by very limited forms of competition; these are sometimes described by critics as "shared monopolies" (Bowles, Edwards, and Roosevelt 2005, 265). In the middle are partial oligopolies, in which large firms may have some control over

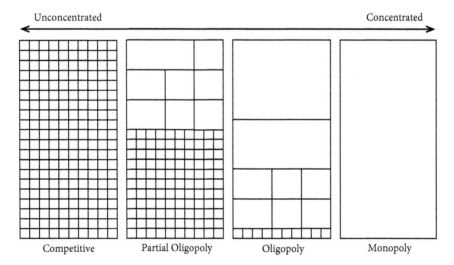

Figure 1.1 Levels of market concentration.
Each rectangle represents a single, hypothetical firm, with size proportional to market share.

their prices but lack the power to significantly reduce competition, as is the case with oligopolies. The word monopoly is derived from the Greek words for single (mono) and seller (polein), but concentrated markets may also occur among buyers, as found in a monopsony or oligopsony (derived from the Greek word for buy, opsōnía).

There has been a strong tendency in more industrialized countries, including the United States, to move away from competitive markets and toward higher levels of concentration (Heffernan, Hendrickson, and Gronski 1999; Du Boff and Herman 2001). As markets go through this process of consolidation, the average firm size increases, barriers to entry for other firms rise, and the remaining firms have more influence over prices, as well as a greater potential for higher profits. These are some of the widely recognized reasons why markets tend to become less competitive—firms understand the benefits to be gained by expanding their market share, reducing the number of competing firms, and increasing their leverage over the terms of exchange (Foster and McChesney 2012). These trends are most evident in industries such as airlines and telecommunications, in which the US government intervened to encourage greater competition, only for them to eventually return to more oligopolistic structures (Brock 2011).

Potential negative impacts

Governments sometimes intervene when industries reach a high level of concentration because such limited competition may lead to numerous negative outcomes. Market consequences could include consumers paying higher prices, suppliers receiving lower prices, or reduced innovation. Oligopolistic firms have disincentives to reduce profits by investing in research and development, particularly when doing so could lead to lower barriers to entry and increasing competition. In addition, large organizational size can discourage innovation indirectly, via complex bureaucracies that are reluctant to approve new ideas (Brock 2011). Increasing size can be an advantage for reducing prices paid for inputs, however, as smaller suppliers may have fewer alternative buyers for their products and less organizational capacity to negotiate the best possible terms (Calvin et al. 2001).

In order to raise consumer prices, it is not necessary for executives to gather in one room and conspire to achieve these markups. When just a few firms control a large share of the market, they can simply indicate their intention to raise prices, and the others will benefit by following suit, a strategy that is called *price signaling* (Baran and Sweezy 1966). Oligopolistic firms can more easily pressure each other to avoid price wars that would lower their profits. The result is an unwritten rule that rivalries based on advertising, product differentiation, and reducing labor costs are expected, but competing on price is unacceptable. John Bell, a former CEO of a coffee company that was eventually acquired by Kraft, explained that he was constantly reinforcing this message to rivals through his speeches and media interviews in order to prevent "the natural competitive reaction." Lowering prices was given the code word "non-strategic" in public communications, and emphasizing his opposition to it was "an indirect means of telling competition to 'play ball' (Bell 2012)."

Nevertheless, executives in industries controlled by a small number of firms may go beyond signaling and are occasionally even caught conspiring to fix prices. A few examples include:

- The US Federal Trade Commission (FTC) discovered in the mid-1960s that leading bakeries and retailers in Washington State had met frequently and agreed to increase the price of bread by 15–20 percent (Parker 1976).

- In 2013, Hershey pleaded guilty to working with other large chocolate manufacturers (Mars, Nestlé, and Cadbury) to raise prices in Canada. While the other firms denied the charges, together they paid more than $22 million to settle a resulting class-action lawsuit (Culliney 2013).

- Western European governments have reported numerous schemes to raise food prices in recent years (e.g., beer, flour, bananas, chocolate and dairy products), which have resulted in fines totaling hundreds of millions of dollars.

As mentioned above, additional problems may result from increasing concentration in the food industry, including negative impacts on communities, labor, human health, animal welfare and the environment (economists define many of these impacts as "externalities").

Regardless of the consequences, as industries consolidate, fewer and fewer people have the power to make important decisions, such as what is produced, how it is produced, and who has access to these products (Heffernan, Hendrickson, and Gronski 1999). Dominant firms are typically controlled by a small board of directors (eleven people on average), with decision-making power concentrated in the hands of the chief executive officer (CEO). The individuals who serve on these boards typically do not reflect the composition of society and in the United States are most often men, of European ethnicity, with an average age of 60 and educated at elite institutions like Ivy League schools (Lyson and Raymer 2000). They also tend to share very similar conservative worldviews through frequent socializing with other members of the upper class in exclusive clubs, summer resorts, philanthropic organizations, etc. (Domhoff 2014).

Unanswered questions

Despite its adverse impacts on society at large, increasing market concentration is frequently portrayed as unstoppable. Business historian Louis Galambos, for example, claims that, "Global oligopolies are as inevitable as the sunrise" (Zachary 1999). Scientists who study social networks, including business networks, have pointed to a tendency for the "rich to get richer" as their initial advantages are magnified and eventually snowball, leading to even greater success (Barabási and Bonabeau 2003; Easley and Kleinberg 2010). Yet these trends have been highly uneven temporally, geographically, and by position within the food system (Friedland 2004). The farming of commodity crops in the Midwestern United States or the regional distribution of many specialized foods, for example, can still be characterized as competitive markets, suggesting that establishing and maintaining an oligopoly is not always as easy as claimed (Lewontin and Berlan 1986; Lewontin 2000).

The interesting question, then, is why have some segments of the food system already become oligopolies and not others? Why don't they all resemble the global soft drink industry, which has been dominated by just two firms, Coca-Cola and Pepsi, for decades? This suggests additional questions, such as what specific factors currently constrain other food and agriculture industries from approaching this organizational model? What factors might enable them? The answers are important because they might help slow, or even reverse, current trends toward concentration. Even if they don't take us quite that far, a better understanding of how concentration occurs could help us to shape its direction and ameliorate some of its negative impacts.

Before answering these questions, however, a critical first step is simply to characterize the changes that have taken place, which can be more difficult than it sounds. Determining the level of concentration in different food industries is complicated by the inaccessibility of accurate sales data (Heffernan, Hendrickson, and Gronski 1999; Fernandez-Cornejo and Just 2007), even for "publicly" held firms. Uncovering just who owns what is also challenged by the opaque and constantly shifting corporate parentage of many brands and subsidiaries. Due to increasing public concern about food issues—evidenced by bestselling books by Eric Schlosser, Michael Pollan, and Barbara Kingsolver, and numerous documentary films that critique the current food system—some firms are taking increasing measures to hide their dominance. Trade journals are less likely to report market shares for leading food and agricultural firms, for example, and the types of acquisitions that typically generated press releases in the late 1990s may now go unannounced.

Differing perspectives

Economics

Industry concentration is studied primarily by economists. One simple indicator that they developed to characterize the competitiveness of markets is a concentration ratio or the sum of the market shares of the top firms. A frequently used number is the top four, sometimes abbreviated as a CR4. Thresholds vary depending on the analyst, but most institutional economists suggest that when four firms control more than 40 percent or 50 percent of a market, it is no longer competitive (Scherer and Ross 1990; Shepherd and Shepherd 2004). A more recent measure, used by regulators such as the US Department of Justice (DOJ) to evaluate mergers and acquisitions, is the Herfindahl-Hirschman Index (HHI). The index is the sum of squares of market share for all the firms in a given market. If one firm, for example, controlled 100 percent of a market, the HHI would be 100 squared, or 10,000. If two firms divided the market, the HHI would be 5,000 (50 squared + 50 squared). The US government once considered markets with an HHI above 1,800 to be highly concentrated (Gould 2010). Notably, this level would not be exceeded if four equally sized firms controlled 80 percent of the market, despite the highly conducive environment for price signaling that would result. In 2010, the DOJ and the FTC raised the threshold even higher, to 2,500—the equivalent of four firms evenly dividing 100 percent of the market. Although the HHI is designed to be more sensitive to changes in market share among the top firms, it is less intuitive than concentration ratios.

A weakness of both measures is that they are designed to character-ize *horizontal* integration within a national (or smaller than national) market. Concentration is increasingly occurring through *vertical* integration, however, as firms buy upstream suppliers or downstream retailers, both in national markets and at the global level. In addition to direct ownership, less formal but still effective means of control are becoming more common, such as strategic alli-ances or contracting arrangements (Heffernan 2000). As a result, the full extent of market power has become much more difficult to establish accurately, and concentration measures may underestimate the ability of firms to enhance their own interests at the expense of others.

Mainstream economists tend to view concentration as unproblematic, due to a strong abstract belief in economies of scale, despite insufficient empirical evidence to support these supposed efficiencies (Johnson and Ruttan 1994; DiLorenzo 1996). Consumers are often claimed to benefit from synergies and lower transaction costs that are expected to result from mergers and acquisitions (Farrell and Shapiro 2001). Because of their organizational complexity, however, many large firms actually encounter diseconomies of scale and experience a loss of efficiency with increasing size (Adams and Brock 2004; Carson 2008). This may explain why acquisitions often fall short of expectations, and an estimated one out of three are eventually undone, via sales to competitors or spin-offs into new firms (Buono and Bowditch 1989).

Most economists also tend to focus on narrow criteria of economic power such as pricing or output measures and ignore other sources of power that can be utilized by large organizations. Walter Adams and James W. Brock (2004, 8), two economists who were critical of this tendency within their discipline, noted that additional powers include:

> the capacity to obstruct technological advance; to manipulate the alternatives from which society is allowed to choose; to coerce society to accede to its demands through threats to shut down facilities or to relocate them elsewhere; to infiltrate government agencies with influential decision makers drawn from the industries ostensibly being regulated; and to obtain government bailouts when collapsing giants are considered to be too big and too important to be allowed to fail.

Political economy

Political economy takes a broader view than economics, recognizing a much higher degree of interaction between governmental agencies and private economic organizations. This field draws on additional disciplines including sociology, political science, geography, and cultural studies. Although it is very diverse, many of its strands are more critical of the status quo than orthodox economics. Political economists are therefore likely to question why markets are organized in their current form, who played a role in this, and how their structure benefits some more than others (Lipschutz 2010). One consequence is a much greater emphasis on power, although this is a difficult concept to define precisely. A broad definition, slightly modified from one proposed by Bertrand Russell in 1938, is "the capacity of some persons to produce intended and foreseen effects on others" (Wrong 1995, 2). Importantly, this capacity can be "naturalized" by institutions, so that the majority of people take it for granted and do not question it (Gramsci 1971; Gibson-Graham 2006).

A strand of political economy that focuses heavily on concentration is Paul Baran and Paul Sweezy's (1966) theory of monopoly capital, which challenges conventional economists' abstract emphasis on so-called perfect or free competition and their lack of attention to the rising number of oligopolies (Box 1.1). This theory has been expanded by colleagues to explore capitalists' impacts on labor (Braverman 1998) and their growing emphasis on speculative finance (Magdoff and Sweezy 1987; Foster and McChesney 2012). Although strongly influenced by Karl Marx, their work has interesting parallels with libertarian political economists, who emphasize the essential role of government regulations and subsidies in facilitating concentration (Stromberg 2001; Carson 2007).

A political economy approach that is frequently employed in food studies is value chain analysis or closely related approaches such as commodity systems analysis (Friedland 1984; Friedland 2004). These typically follow a single product, from its design and production to its consumption, to understand how the entire system works. Value chain analysis has become much more global, as more food and agricultural firms have expanded around the world in search of new markets and lower material and labor costs (Dixon 2002; Pritchard and Burch 2003). Although the general approach is also applied in economics and business, critical political economists place more emphasis on analyzing differences in power in the relationships between firms and individuals throughout the chain. Most commodity systems and value chain analyses of foods have heavily emphasized the production stages and particularly the role of labor. There is an increasing recognition of the need to place more emphasis on consumption, however (Lockie and Kitto 2000; Goodman 2002; Goodman and DuPuis 2002).

Karl Polanyi's *The Great Transformation* (1944) has also been influential among those who study food. Polanyi suggests that there is a "double movement" that results when the negative impacts of capitalist expansion incite a spontaneous, defensive reaction. This helps to explain a number of movements against the dominant food and agricultural system, from US farmer protests against railroads in the late 1800s (Constance, Hendrickson, and Howard 2014) to global certification of fair trade labels beginning in the late 1980s (Jaffee 2007). The theory can be criticized for being "underspecified" and failing to predict exactly what would trigger such a response, however (Munck 2006, 185).

"Counter-movements" have certainly not proven to be automatic, especially if the impacts of capitalists' actions are hidden or legitimized, or if government repression is successful.

Box 1.1 The Missouri School of Agriculture and Food Studies

Although all of my grandparents were raised on farms and grew much of their own food, my childhood was spent in the suburbs. Virtually all of my food came from the supermarket, and I knew very little about the system that produced it. In the late 1990s, when I became a graduate student at the University of Missouri, I was quite surprised to learn of the increasing levels of concentration in US food industries, and how powerful a small number of firms had become.

This was a major focus of what has become known as the "Missouri School" of agricultural and food studies (although it has had far less influence than the Chicago School of economics that originated at the University of Chicago in the 1940s). The group developed as a result of the efforts of William D. Heffernan, along with students and colleagues at the University of Missouri's Department of Rural Sociology (Kleiner and Green 2009). Heffernan studied the dynamics of the poultry industry beginning in the 1960s, focusing on the increasing use of contracts between processors and producers. He found that this had some initial advantages for poultry growers, but power shifted dramatically toward processors over time (Heffernan 1972; Hendrickson et al. 2008a). Later, he and his co-authors examined concentration and its impacts on local communities for other Midwestern commodities, such as beef, pork, corn, and soybeans (Heffernan, Hendrickson, and Gronski 1999). My first collaboration with these researchers involved a study of structural changes in the retail and dairy industries (Hendrickson et al. 2001).

The approach of the Missouri School is pragmatic, with an emphasis on characterizing problems in the food system and assisting affected communities to address them (Constance et al. 2014). Rather than adhering to one specific theory, its practitioners draw on multiple perspectives, including those of Max Weber, Karl Marx, and Thorstein Veblen, as well as more recent approaches, such as those of monopoly capital theorists associated with the *Monthly Review* (Baran and Sweezy 1966; Foster and McChesney 2012). Bonanno (2009) notes that the Missouri School is most closely aligned with the philosophy of John Dewey, which is fundamentally anti-elitist. This means that we would oppose concentration even if it did not lead to any negative impacts on society or the environment, as it gives a small minority great power over the food we eat.

Capital as power

In this book, I employ a political economy approach but draw more specifically on Jonathan Nitzan and Shimson Bichler's theory of *Capital as Power* (2009). This theory seeks to understand capital from the point of view of capitalists, especially those who benefit the most from the current system: the largest corporations and the wealthiest individuals, who are typically major shareholders in these firms. Capital as Power recognizes that corporations quantify their perceived influence through "capitalization." Technically this is calculated as the firm's current share price multiplied by the number of shares outstanding, but it can be viewed as an estimate of the future stream of earnings in present values while adjusting for perceived risks. Another way of thinking about capitalization is that it is a quantification of capitalists' consensus expectations that people will continue to acquiesce to the firm's power. It therefore measures not just a firm's capacity to provide goods and services but "the power of its owners and directors to shape and reshape politics, society and culture" (Di Muzio 2013, 6).

Echoing O'Brien in the dystopian film, *Nineteen Eighty-Four*, they note that power is not only a means of accumulation but "also the *ultimate end* of accumulation" (Nitzan and Bichler 2009, 16). Capitalism as a system is therefore better understood as a mode of power rather than a mode of production. This shift, in comparison to other theories, places even more emphasis on social relations than on material objects (e.g., embodied labor or utility). It also seeks to understand both quantitative changes in markets and qualitative changes in society as part of the same process of the accumulation of power (Baines 2015).

An important aspect of this perspective is the view that dominant firms do not try to maximize profits, as is typically argued by other political economy perspectives (see Magdoff and Foster 2011). Instead, firms compare their performance to close competitors and use benchmarks, such as the S&P 500 share price index, to monitor the average capitalization for the largest firms. Differential accumulation may not sound substantially different than profit maximization, but it is a far more realistic description of corporate behavior and with some very important implications. In periods of stagnation, for example, growth rates may slow, but dominant capitalists are content to grow faster than the average or even to decline less than the average, as this means their capitalization and power are still increasing in relative terms. Conditions of high economic growth are actually riskier, because capitalists are more likely to lose power relative to others (Nitzan and Bichler 2014). Additional empirical support for this view comes from Thomas Piketty's (2014) analysis of several centuries of income and wealth data in developed countries, which found increasing inequality during periods of low growth rates (Piketty 2014).

Conflict drives this competition, which means that capitalists have little choice but to try to increase their income and assets relative to those of others, or risk going out of business, sometimes via acquisition by a competitor (Bichler and Nitzan 2014). This requires active efforts to restructure markets and society in ways that increase their power, including encouraging increased consumption of their products and sabotaging potential alternatives, particularly those that would allow people to be more self-reliant. All of these actions focus on increasing profits but *in comparison to other firms*, not to achieve a theoretical maximum. An intended increase in inequality is one result, as well as numerous additional negative impacts, which, from the perspective of capitalists, could be viewed as collateral damage (Cochrane 2010).

The Capital as Power approach challenges some widely held categorizations. Many political economists, such as the monopoly capital group, view finance as "fictitious" or separate from material capital, such as physical plants and equipment (Hager 2013, 43). Nitzan & Bichler, in contrast, state that when the object of accumulation is viewed as power, "all capital is finance, and only finance (2009, 262)." They also claim there is no distinction between dominant capital and governments, suggesting that their interests overlap to such a great extent and the power of capitalists is so dependent on government actions that boundaries are meaningless (Nitzan and Bichler 2009). Sympathetic critics have pointed out, however, that governments are subject to other logics, in addition to accumulation (Starrs 2013). James O'Connor (1973), for instance, described the tension between accumulation and legitimation, or the need for policies to be seen as justified by the majority of the population. A government's legitimacy can be undermined if the public recognizes harms resulting from policies that aid dominant capitalists' strategies of accumulation. Therefore, when resistance threatens to undermine the stability of the system, governments take at least symbolic actions to maintain an appearance of civic interest (Green and Heffernan 1984).

Concentration is a key strategy to increase power, one that Nitzan and Bichler (2009) call *breadth*. This path involves either internal growth (also called greenfield or organic growth) that is faster than competitors or external growth via mergers and acquisitions. External growth is often preferred because it is a less risky means of increasing size and power. Even if such combinations fail to increase, or even reduce, productive capacity or potential, they frequently result in a higher total capitalization—reflecting an expected increase in power (Nitzan 1998). Acquisitions are far more common than mergers, as firms that are already dominant are more likely to have internally-generated resources or can borrow to finance buyouts of other firms.

The other key strategy is described as *depth*, which involves cost cutting or price increases (Nitzan and Bichler 2009). As discussed above, firms that are more successful with breadth, and therefore achieve greater market share, also gain significantly more power to enact price increases. Nitzan and Bichler suggest that cost-cutting strategies are more easily replicated by rival firms, and thus less effective for outperforming benchmarks, but this does not give enough attention to the possibility that the largest firms have more power to (1) demand lower prices from suppliers, (2) negotiate lower wages for workers, or (3) obtain government subsidies.

Power in the food system

Because we depend on food to live, concentration raises more concern in food industries than in most other economic sectors. The importance of food also makes it a key site of contestation in economic, political, and cultural realms. Virtually all governments have thus adopted special agricultural policies—for example, to ensure that food is produced in sufficient quantity and available at affordable prices while keeping farmers economically viable (Hendrickson et al. 2008b).

Which firms are the most dominant in the global food system? Table 1.2 lists those engaged in food and agriculture that are among the top 500 firms in the world according to market capitalization. Most of these focus on packaged foods (eighteen firms, $1.46 trillion), followed by retail (eleven firms, $685 billion) and agricultural inputs (nine firms, $557 billion). Commodity firms and distributors are represented by just one firm in each case, near the bottom of the list, while farming is a segment of the food system that does not have any firms of this magnitude. The result is an hourglass-shaped system, with a large number of farmers at the top, an even larger number of people who eat food at the bottom, but a much smaller number of firms in the middle that control how food is moved from producers to consumers (Heffernan, Hendrickson, and Gronski 1999).

Adopting a broad understanding of power, the following chapters explore how it is exercised in each major stage of the food system, as well as the organic food system—organic originated as an alternative to the mainstream but is an increasingly consolidating industry. With this value chain approach, I also focus on a more limited number of commodities/foods, such as soybeans, pork, milk, and leafy greens across some of these stages. The emphasis is on the United States, as nearly half of the firms in Table 1.2 are headquartered in

Table 1.2 Global market capitalization of dominant food firms, 2014

Firm (headquarters if not the United States)	Market capitalization in billions ($)	Primary stage of food system
Walmart	247	Retail
Nestlé (Switzerland)	243	Packaged Foods & Beverages
Coca-Cola	170	Packaged Foods & Beverages
Anheuser-Busch InBev (Belgium)	169	Packaged Foods & Beverages
PepsiCo	128	Packaged Foods & Beverages
Unilever (Netherlands)	119	Packaged Foods & Beverages
Ambev (Brazil—subsidiary of AB InBev)	118	Packaged Foods & Beverages
Bayer (Germany)	112	Agricultural Inputs
BASF (Germany)	102	Agricultural Inputs
McDonald's	97	Retail
SABMiller (UK)	80	Packaged Foods & Beverages
Diageo (UK)	78	Packaged Foods & Beverages
Caterpillar	63	Agricultural Inputs
DuPont	61	Agricultural Inputs
Monsanto	60	Agricultural Inputs
Dow	59	Agricultural Inputs
Mondelez	59	Packaged Foods & Beverages
Starbucks	55	Retail
Costco	49	Retail
Danone (France)	45	Packaged Foods & Beverages
Woolworths (Australia)	42	Retail
Heineken (Netherlands)	40	Packaged Foods & Beverages
Tesco (UK)	40	Retail
Target	38	Retail
Associated British Foods (UK)	37	Packaged Foods & Beverages
Syngenta (Switzerland)	35	Agricultural Inputs

(continued)

Firm (headquarters if not the United States)	Market capitalization in billions ($)	Primary stage of food system
Seven & I (Japan)	34	Retail
Deere	34	Agricultural Inputs
Kraft	33	Packaged Foods & Beverages
Yum! Brands	33	Retail
General Mills	32	Packaged Foods & Beverages
Femsa (Mexico)	32	Packaged Foods & Beverages
Pernod-Ricard (France)	31	Packaged Foods & Beverages
PotashCorp (Canada)	31	Agricultural Inputs
Archer Daniels Midland	28	Commodities
Carrefour (France)	28	Retail
Kweichow Moutai (China)	26	Packaged Foods & Beverages
Kellogg	23	Packaged Foods & Beverages
Kroger	22	Retail
Sysco*	21	Distribution

Source: Financial Times 2014. Note that many firms include sales of products or services unrelated to food and agriculture or extend into multiple stages of the food system through vertical integration. *Sysco was in the top 500 globally in 2013 but fell just below this threshold in 2014.

this country. In addition, market shares and concentration ratios, which provide quantitative indicators of their power, are more easily available at a national level. Comparisons are frequently made with other regions, however, particularly when these dominant firms extend their influence to other parts of the world.

Each chapter also focuses on a key qualitative strategy that these firms use to restructure society, overcome restraints on concentration, and increase their control. Although they make use of numerous strategies, an in-depth examination of specific techniques, and the resistance they often provoke, challenges the notion that current levels of concentration were inevitable. Instead, it details the enormous efforts that firms must expend in order to continually increase their power, using strategies that include:

- changing the interpretation of antitrust laws (Chapter 2)
- structuring exchange networks (Chapter 3)

- reshaping consumption habits (Chapter 4)

- manipulating prices (Chapter 5)

- maintaining government subsidies (Chapter 6)

- strengthening intellectual property protections (Chapter 7)

- influencing voluntary standards (Chapter 8)

In addition to the approaches noted above, a number of additional strategies are also briefly explored in boxes in subsequent chapters. Most of these boxes focus on the actions of second-ranked firms in food and agricultural industries, which typically receive less attention from researchers than the top firms, such as Walmart, AB InBev, and Archer Daniels Midland. Although the leading firms are likely to have the most disproportionate influence on society, a lower-ranked firm's choice of strategies may succeed in achieving a shift in power, quantified by overtaking the top position.

The pattern that emerges is that capitalists increase their influence on society by being extremely flexible. Because their initiatives are rarely unopposed, achieving firm and industry goals requires close cooperation with allied organizations and reacting quickly to potential threats to their success, even when they appear insignificant. They are often able to circumvent challenges, adapt to demands for change, and co-opt potential forms of resistance. Their power is frequently hidden by being exercised indirectly, such as through their influence on government regulations and enforcement, on upstream or downstream firms, and key organizations (e.g., the mass media, foundations, think tanks, and universities). An examination of these dynamics, however, indicates that capitalists can sometimes be pressured into ameliorating some of their negative impacts, especially when the cost of not doing so would threaten the foundations of the system that gives them so much control.

The next chapter describes how capitalists responded to one such challenge in the last century. At that time, the enactment of antitrust regulations helped restore public confidence in a political economic system that was becoming dominated by a small number of firms. These regulations are now being reshaped to reduce these limits to their power. Retailing is one stage of the food system that, as a result, has become much more concentrated in recent decades.

Chapter 2

Reinterpreting antitrust: retailing

Now all restaurants are Taco Bell.
—Lenina Huxley (Demolition Man)

Retailing is the closest link in the food chain to consumers. This structural position gives these firms a gatekeeper role and thereby the potential to wield enormous power over both consumers and suppliers. This potential wasn't effectively realized in the United States until recently, however, particularly beyond the fast food sector. The grocery industry, for example, stood out in comparison to its counterparts in other industrialized countries for its relatively fragmented structure until the late 1990s (Wrigley 2001). Rigorous enforcement of US antitrust laws encouraged supermarket chains to instead exercise dominance at a more limited regional level through the 1970s. Changes in judicial interpretation and enforcement since then, however, have allowed firms to become increasingly national and global in scope.

This chapter explores how three different types of food retailers have benefited from government relaxation of antitrust regulations in recent decades: supermarkets, convenience stores, and fast food restaurants. It also examines the negative impacts of these changing power relations on society, particularly driving down prices for suppliers, lowering wages for retail workers, and limiting choices for consumers. The potential to revitalize antitrust enforcement is then analyzed, using examples of key retailer actions that have experienced opposition from regulators.

The shifting landscape of antitrust regulation

Antitrust regulation has a long and complicated history. As long ago as the late 1800s, the negative impacts of industry concentration led to social movements opposing mergers and acquisitions. Many of these movements were led by farmers, who were exploited by powerful banks, railroads, and meat processors (Goodwyn 1978). The radical Populist Movement in the United States and Canada pressured national governments to enact antitrust laws, such as the Anti-Combines Act in Canada in 1889, and the Sherman Antitrust Act in the United States in 1890 (Scherer and Ross 1990). The reformist Progressive Movement in the United States also worked to counter the growing power of trusts and successfully lobbied in favor of legislation, including the Clayton Antitrust Act of 1914, the Federal Trade Commission Act of 1914, the Packers and Stockyards Act of 1921, and the Robinson-Patman Act of 1936. Paradoxically, these laws neutralized much of the opposition to trusts by legitimizing large corporations and steering societal conflicts into the legal-technical arena; they actually reduced competition in many industries, such as railroads, as a result (Kolko 1963; Neuman 1998). Although somewhat selectively enforced, the legislation nevertheless did result in the breakup of some trusts, such as the meat trust, and hindered concentration in others.

The Robinson-Patman Act was particularly relevant to retailers, as it was also known as the "Anti-A&P Act" (Lynn 2006). The Great Atlantic & Pacific Tea Company, or A&P, was the first supermarket chain in the US, and it was the Walmart of its day, operating 16,000 stores by 1930 and controlling 12 percent of the national grocery market (Wrigley 2001). The firm used its size to negotiate discounts based on volume from suppliers. Smaller competitors were unable to obtain these cost reductions, and their higher retail prices led consumers to shift their purchases to A&P. The Robinson-Patman Act prohibited these price breaks, due to their anti-competitive impacts and the potential for A&P to become even more dominant in the grocery industry. A sponsor of the legislation, Wright Patman, stated: "The expressed purpose of the Act is to protect the independent merchant and the manufacturer from whom he buys" (Lynn 2006).

By the 1970s, however, dominant firms in numerous industries were chafing against the limits imposed by antitrust laws, particularly in their efforts to increase their power faster than competitors. They employed numerous strategies to overcome these limits but one that was quite successful was contributing financially to conservative politicians who campaigned on platforms of deregulation. In 1980, Ronald Reagan, a Republican from California, was elected president of the US. He directed regulatory agencies to reduce their enforcement of antitrust laws, with the rationale that large firms were now competing in a global market. Deals that would not have been allowed previously were swiftly approved at the national level. In addition, the burden of proof on those potentially harmed by concentration was raised (Potts 2011).

Another very effective strategy was to influence judges in order to change the way they interpreted antitrust laws. Beginning in the late 1970s, large corporations funded public and private think tanks, which in turn arranged for judges to go on all-expenses-paid junkets. These were typically held at resorts in warm weather states, such as Florida and Arizona, where the judges could play golf. While there, they would also attend seminars presented by Chicago School economists suggesting that mergers and acquisitions would increase efficiency, and should not be opposed unless there was clear evidence of harm to consumers. This was a dramatic change from the intent of legislation noted above, and most prominently advocated by University of Chicago graduate Robert Bork, who was a law professor at Yale (Olson 2014). By the early 1990s, an estimated two-thirds of all federal judges had participated in at least one of these programs, which was affiliated with George Mason University (Aron, Moulton, and Owens 1994). A judge even stated in this program's promotional literature that, "as a result of my better understanding of marginal costs, I have recently set aside a $15 million anti-trust verdict" (ibid.). Ruling against the plaintiffs in anti-trust suits is increasingly the norm, and while the laws still technically exist, they have essentially been repealed by judicial interpretation. Although retailers were far from the only firms to play an active role in reshaping the regulatory landscape, they were able to take advantage of these changes.

From regional to global dominance: supermarkets

By the late 1990s, the relaxation of antitrust regulations fueled a wave of consolidation throughout the entire US economy. Supermarket chains were no exception, and these firms dramatically increased their merger and acquisition activity. Some additional factors that contributed included: (1) easy access to financing (2) low rates of inflation that inhibited price increases as a means of increasing profits, and (3) slower growth in the markets of industrialized countries (Wrigley 2001). Perhaps one of the strongest motivators for supermarkets, however, was the entrance of the world's largest retailer, Walmart, into the grocery industry, which encouraged takeovers by its competitors as a defensive strategy.

Before Walmart entered the food business in 1988, the largest US supermarkets had very high market shares in specific metropolitan areas or even states, but none had a fully national presence. With Walmart changing this norm, the fastest way for other firms to respond was to buy their way into other regions. The USDA's Phil Kaufman notes that between 1997 and 2000, 4,100 stores were acquired, which totaled approximately 20 percent of all U.S. supermarkets

(2002, 26). As a result, the CR4 for grocery sales increased dramatically, from 17 percent in 1987 to 27 percent in 1999 (Calvin et al. 2001).

Acquisitions also occurred in other regions of the world, including Europe, despite already high levels of concentration—for fifteen EU countries, the top five firms controlled a weighted average of 49 percent of the market in 1999 (Dobson, Waterson, and Davies 2003). In the United States, Walmart was able to convert its existing general merchandise stores into "supercenters" that also sold food, but in other parts of the world, buyouts were a faster route to expansion for this firm as well. Walmart acquired retail chains in Germany and the UK, for example, which quickly encouraged the merger of French retailing giants Carrefour and Promodes (Hendrickson et al. 2001). Walmart, along with Carrefour, Tesco (UK), and Metro (Germany) also made acquisitions or formed joint ventures in less industrialized countries—particularly in Asia but including parts of Latin America and Africa—to become truly global food retailers (Wilkinson 2009).

Walmart has become the face of globalization, due to the scope of both its retail operations and its supply chains. In just a few decades, the firm grew from a single store in Arkansas to become the largest corporation in the world (Belsie 2002). The firm's annual sales exceed the Gross Domestic Product of all but 23 nations (Howard 2014a). In the US, Walmart controlled approximately one-third of grocery sales by 2011, more than three times the market share of its closest competitor (Table 2.1). These competitors continue to make acquisitions in an attempt to keep up, such as Kroger's acquisition of the Harris-Teeter chain in 2014, and Safeway's combination with Albertson's that same year. The market share of dominant firms continues to be much higher when measured in smaller geographic regions, with concentration ratios exceeding 80 percent in 231 metropolitan areas (Food & Water Watch 2013).

One key impact of this supermarket concentration has been to shift the balance of power away from suppliers. Wall Street analyst Mark Husson

Table 2.1 US grocery market, 2011

Firm	Market share (percent)
Walmart	33
Kroger	9
Safeway	5
Supervalu	4
	CR4: 51

Source: Clifford (2011).

explained that food manufacturers once "ruled over retailers the way Romans did with their tribes. It was divide and conquer. But in the last two or three years, the tribes have gotten together. They're now dictating price and promotions to the manufacturers" (Hendrickson et al. 2001, 718). This shift has increased the influence of retailers on other upstream stages of food system as well, such as distribution and farming/ranching.

Walmart developed more efficient supply chains than competitors in general merchandise and applied these lessons to the retail food industry in the 1990s, which were quickly copied by competitors (Baines 2014a). These dominant firms effectively pressured their suppliers to provide discounts based on volume, demonstrating how irrelevant the intent of the Robinson-Patman Act had become by this time. Walmart now accounts for a significant percentage of sales for many of the world's largest manufacturers, frequently 20 percent or more, and the retailer utilizes its importance as the largest customer to demand increasingly lower prices (Olson 2014). The journalist Charles Fishman (2003) described this dynamic in detail, illustrating how the retailer was able to sell a twelve-pound, one-gallon jar of Vlasic pickles for just $2.97 in the late 1990s. Vlasic declared bankruptcy soon after this concession, although it was not the only factor that contributed. Walmart implicitly threatens suppliers that they can source from competitors in China and other low wage nations, which has led many manufacturers to shift their production to these sites in order to meet the retailer's price demands (Fishman 2003).

Walmart's low prices reflect not only this ability to dominate suppliers but numerous hidden subsidies, such as cheap transportation, government assistance for their low paid laborers, local tax incentives, and federal tax avoidance strategies (Americans for Tax Fairness 2014). Smaller retailers are less likely to benefit from these subsidies and, as with A&P's competitors in the early 1900s, many have gone out of business. One estimate suggests that approximately half of the decline in small retailers in the 1990s was due to Walmart's expansion (Jia 2008). Such low retail prices have also discouraged new entrants into what became a much more capital-intensive industry (Wrigley 2001).

Labor relations have also been strongly affected by recent changes in the supermarket industry. Walmart is famously anti-union, and the firm has gone as far as closing stores that successfully unionized, such as one in Quebec in 2005 (Palmer and Martell 2014). The firm's competitors have sought to obtain the same advantage of low wages, which has resulted in numerous labor disputes. Beginning in 2003, Safeway experienced the longest grocery strike in history when workers refused to accept cuts to their wages and benefits. At the time the CEO, Steven Burd, received compensation of more than $11 million per year.

The standoff lasted nineteen weeks before the union caved in to the corporation's demands. The settlement signified an industry-wide shift from near middle class wages to low income wages—the proportion of California food retail workers receiving poverty level wages increased from 43 percent in 1999 to 54 percent in 2010, and more than 40 percent of these workers are Latino (Jayaraman 2014).

Safeway's "competitors," Albertson's and Kroger, colluded with the corporation to lock out their own employees, who were members of the same union, within 48 hours after Safeway workers went on strike. This affected a total of 59,000 workers. The three firms also shared costs and revenues in order to survive the loss of business they experienced during the strike, and to successfully break the union. The attorney general of California filed an antitrust suit against the firms as result of these actions but lost in the initial district court, as well as upon appeal (California v. Safeway Inc. 2011).

Despite the industry's attention to reducing costs from labor and suppliers, as the industry consolidated, consumers experienced higher prices, with food costs increasing twice as fast as inflation and wages from 2010 to 2012 (Food & Water Watch 2013). Some supermarkets have even been accused of price fixing: two Chicago-area chains were alleged to have colluded on the price of milk between 1996 and 2000, raising the average cost to approximately one dollar more per gallon than other regions. Jewel and Dominick's—later acquired by Supervalu and Safeway, respectively—faced a class action lawsuit as a result, but a judge dismissed this case for insufficient evidence (Quigley 2003).

The negative impacts of concentration have disproportionately affected low income and minority consumers. There is increasing media coverage of what are called "food deserts," for example. This term originated from a resident of a low income housing complex in Scotland, who was describing the lack of healthy food options in the area (Cummins and Macintyre 2002). Public health scientists have largely avoided this term because research indicates there are few heavily populated areas in industrialized countries where food is completely lacking. In addition, food deserts are frequently defined in very simplistic terms, such as areas that do not have chain supermarkets (Walker, Keane, and Burke 2010). After the Second World War, supermarkets tended to expand to suburban areas and gradually moved out of inner city areas, for both economic and cultural reasons. Stores in inner cities areas had higher rents, higher wages and more losses due to theft and vandalism, which were among the factors pushing chains out of these locations (Eisenhauer 2001). In addition to the affluent populations residing in new suburbs, government subsidies—including tax breaks, tax credits, free or reduced cost land, utilities, and roads—often totaled millions of dollars per store and were strong pull factors (Americans for Tax Fairness 2014). By the 2000s, in cities as large as Detroit, there were no chain supermarkets to be found within the city limits, although some independent grocers did sell similar product assortments (Griffioen 2011).

As a result, in many urban areas, particularly those with large ethnic minority and/or low income populations, there is more limited access to affordable, nutritious foods, such as fresh fruits and vegetables, when compared with affluent areas. Conversely, there tends to be much greater access to less nutritious foods, such as soft drinks, salty snacks, and candy (Larson, Story, and Nelson 2009). Studies have found conflicting results with respect to the health impacts of differing food access, but the current weight of evidence suggests that easier access to unhealthy foods may have more severe consequences for diet-related diseases than limited access to healthier foods (Ford and Dzewaltowski 2008). There have been a few studies examining the impact of the opening of supermarket chain in an area that previously lacked these types of retailers, but purchasing and eating patterns did not change as rapidly as expected (Cummins and Macintyre 2006; Cummins, Flint, and Matthews 2014).

Resistance to the negative impacts of dominant supermarket chains has been organized on many fronts, including opposition to the treatment of labor, and critiquing the environmental impacts of constructing massive buildings and parking lots on the edges of metropolitan areas (Ingram, Yue, and Rao 2010). Critics have successfully blocked or postponed hundreds of proposed Walmart stores in the United States, for example, including New York City and Washington, DC (Baines 2014a). One response has been to emphasize the positive benefits of increased scale—Walmart and even Kroger are so large that when they improve their energy efficiency or reduce waste, the impacts are much greater than if numerous smaller competitors took the same actions. Despite Walmart's public relations emphasis on its sustainability efforts, however, nearly all were adopted because they reduced costs and increased leverage over suppliers (Mitchell 2012).

Perhaps the biggest challenge to the industry is not community activists but other types of retailers. The supermarket share of retail food sales has decreased from 76 percent in 1999 to 67 percent a decade later (Wood 2013). The grocery market is becoming more fragmented, with more sales at retail formats, such as warehouse stores, convenience stores, specialty grocers, mass merchandisers, dollar stores, and drugstores (Martinez 2007). These stores appeal to such attributes as low price, quality and convenience that cannot all be met in a large supermarket format, but supermarket chains have also diversified into many of

Box 2.1 Vertical and Horizontal Integration: Kroger

Compared to other dominant supermarket chains, Kroger retains a high degree of vertical integration, meaning it is more directly involved in other stages of the food system, such as manufacturing and processing. The firm has forty-one processing plants and fifteen dairies, for example, and produces many of its own private label products rather than contracting with outside firms (Progressive Grocer 2002). This level of control once posed some problems with respect to antitrust regulations. In 1968, for example, the firm was sued to prevent it from building a planned dairy processing plant—due to the estimated 20 percent or more of supply that it would control in the region surrounding St. Louis, Missouri—although the suit was ultimately dismissed (Hiland Dairy, Inc. v Kroger Company 1968). Until supermarkets became large enough increase their power over external suppliers, however, such vertical integration provided Kroger with an advantage over its competitors.

Although Walmart has increased its market share in the United States entirely by opening new stores, Kroger has used horizontal acquisitions to fuel much of its growth. More recent Kroger acquisitions are depicted in Figure 2.1, which show some of the subsidiary supermarkets now owned by the firm. As this figure illustrates, somewhat unusually in this industry, the firm was not quick to rename the chains it acquired. This "stealth ownership" provided consumers the illusion of numerous competing retailers in some markets, such as Salt Lake City, Utah, where Fred Meyer, Smith's, and Food4Less were all owned by Kroger. Operationally, Kroger has been more decentralized than supermarkets of similar size, giving the regional chains autonomy in making many decisions. In response to the increasing power of Walmart, however, Kroger changed its purchasing operations to be more centralized, in order to exert more power over other suppliers—this program is called Big Buy, Big Sell (Wrigley 2002). Kroger has also diversified by acquiring firms engaged in other retail formats, including department stores, jewelry stores, and several convenience store chains.

these other formats (Box 2.1). Walmart, for example, owns Sam's Club, which had a 38.4 percent share of the warehouse store market in 2012, although the leading firm Costco held a 46.5 percent share (Skariachan 2013). Another threat comes from the online retailer Amazon, which is expanding its direct delivery grocery business after testing it in several large West Coast cities.

While farmers' markets may appear to be too small to pose a challenge to supermarket chains, this type of retailer is growing exponentially in the United

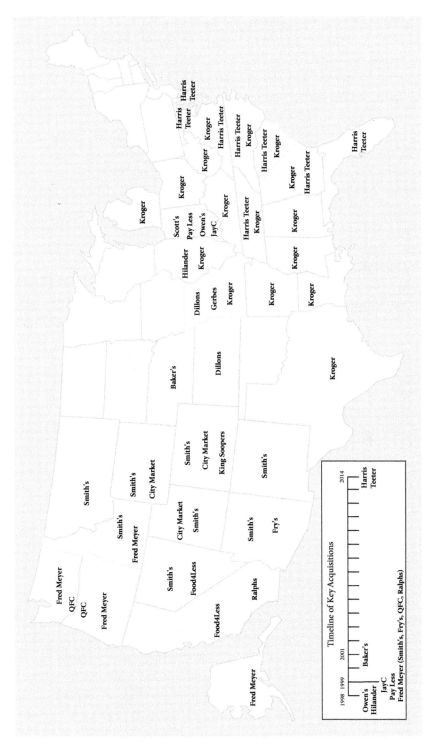

Figure 2.1 Map and timeline of Kroger's acquisitions, 1998–2014.

States: their numbers have increased from less than 2000 in 1994 to more than 8000 in 2013 (USDA 2013). The rules vary, but typically these markets require that farmers engage in selling their own produce, thus giving customers an opportunity to interact and ask questions about their growing practices. In some parts of the country, there is not enough supply from farmers to meet the demand. Supermarket chains have responded to this increasing popularity by setting up farmers' market style displays in their produce sections, such as fruits piled into small wicker baskets and chalk boards listing prices. Safeway and Albertson's even signed their outdoor produce displays in front of the store with the term "farmers market," although a barrage of complaints convinced Safeway to change their signage to "outdoor market" (Wingfield and Worthen 2010). Some smaller chains have allowed free use of their parking lots to host farmer's markets due to the positive publicity and even increased sales this brings to the stores (Burros 2008).

Although dominant supermarkets appear to be maintaining their control, the rapid expansion of power they experienced in the 1990s and early 2000s may be approaching its limits. Global firms, such as Walmart and Tesco, have had less success in beating average returns, such as the S&P 500, since 2003 (Baines 2014a). Walmart is no longer the largest corporation in the world when measured by market capitalization, and by 2014 it had slipped to number ten in the Financial Times Global 500 rankings. Many factors have played a role, but, importantly, the firm experienced setbacks in its international expansion plans, such as acquisitions and joint ventures in Germany, South Korea, and India that were unsuccessful (Berfield 2013). Tesco attempted to expand into the United States in 2007, with a format called Fresh & Easy, but the venture performed worse than expected. It was then sold to an investment firm founded by the billionaire Ron Burkle. Joseph Baines suggests that when compared to grain traders (Chapter 5) and agricultural input firms (Chapter 7), retailers are increasing their power much more slowly (Baines 2014b). Nevertheless, very few firms are larger than Walmart, and the retailer continues to exercise tremendous influence over farmers, food processors, and consumers (Richards et al. 2013).

The supermarket industry also remains quite lucrative for those that have survived consolidation. Compensation exceeds $10 million annually for CEOs of all four leading firms (CompanyPay.com 2014). The six heirs of the founder of Walmart receive billions of dollars per year in dividend income (Banjo 2012). They also increased their wealth to the equivalent of the bottom 41.5 percent of US families in 2010, up from the equivalent of 30.5 percent just three years earlier (Allegretto 2012). The Walmart heirs continue to spend millions of dollars on lobbying to try to repeal estate taxes, despite already avoiding more than

$3 billion dollars in such taxes through loopholes involving complicated trusts (Americans for Tax Fairness 2014).

Cornering the market: convenience stores

Although the supermarket share of grocery sales has declined, convenience stores sales increased 12 percent from 2002 to 2012 (Wood 2013). These retailers tend to have higher prices and much more limited offerings than supermarkets but are within more accessible distances for most of the population. Their very small store size and higher markups create lower barriers to entry compared to other food retail formats. Although the industry is becoming more concentrated, it remains more competitive than other retail sectors. Its structure is characterized by a large number of independent firms, including 63 percent with just one store—although one chain, 7-Eleven, has become much more dominant than its competitors (ibid.).

7-Eleven, and its Japanese parent corporation, Seven & I Holdings Co., controlled an estimated 24 percent market share in the United States in 2012 (Table 2.2), with plans to grow via more acquisitions (Yamaguchi 2013). In 2012 alone the firm made nine major acquisitions, resulting in more than 600 additional stores (Ruben 2013). The firm has undergone numerous ownership changes since its founding in 1927, which resulted in many shifts in strategic direction. In contrast to the current aggressive acquisition strategy, in the late 1990s the firm sold stores in parts of the Midwest and Southeast to competitors, such as Kum & Go, headquartered in West Des Moines, Iowa. The current owners plan to restructure the US industry to resemble Japan, where 7-Eleven has an even larger share of the convenience market (32 percent in 2012), along with 10 percent of all grocery sales (Yamaguchi 2013). The firm is also expanding globally, with outlets in North America, Scandinavia, Australia, and large parts of Asia.

One key to its continued growth was its successful defense against a number of antitrust lawsuits over the years, filed by its own franchisees. The firm uses a

Table 2.2 US convenience store market, 2012

Firm	Market share (percent)
7-Eleven	24
Circle K	1.6
The Pantry, Inc.	0.1
	CR3: 25.7

Source: Yamaguchi (2013).

franchise model to operate many of its stores, which means that they lease their business model to entrepreneurs and do not maintain direct ownership of these outlets. Many of the disputes center around the high degree of control Seven & I has over franchisees, such as dictating retail prices and requiring participation in the firm's payroll system—this is closer to an employer-employee model than a franchise. Attorneys note that although antitrust laws once gave franchisees power to challenge such abuses, these lawsuits have rarely succeeded in recent years (Sparks 2012). 7-Eleven has also been accused of discriminating against Southeast Asian immigrants, who make up a significant proportion of their franchisees. A 2014 lawsuit claimed the firm targeted these ethnic groups in order to terminate franchise agreements and acquire stores at little cost to the parent corporation (Heller 2014).

The second place spot is held by Circle K and its Canadian parent corporation (Alimentation Couche-Tard Inc.), with just 1.6 percent of sales. This market share is likely to increase rapidly, however, as the firm continues its strategy of making numerous acquisitions. Couche-Tard was founded in 1980 by Alain Bouchard, who consolidated Canada's convenience store industry, and is now implementing the same strategy in the United States. The firm views every competitor except 7-Eleven as a potential acquisition (Preville 2014). Couche-Tard attempted a hostile takeover of a Midwestern chain called Casey's for $2 billion in 2010 but did not succeed. 7-Eleven also made an offer to buy the chain, but Casey's decided to remain independent at the time. Although not as global as Seven & I, Couche-Tard has expanded into Europe through acquisitions, and analysts suggest it is likely to expand into Asia next (Ratner 2014).

The pace of acquisitions in the United States increased in 2011, and the prices paid for buyouts grew well above standard industry valuations, suggesting that concentration would continue to accelerate. Larger retailers, such as supermarket chains, are also diversifying into convenience stores, which is likely to further consolidate the industry. Kroger is the most notable example, but other retailers are experimenting with this format, such as *Walmart to Go* and *Safeway X-press* (Souza 2014). In the absence of strong enforcement of the Robinson-Patman Act, these firms should be able to leverage their greater power over suppliers to undercut the prices of competitors, just as they have in the supermarket industry.

Areas that lack chain supermarkets often have an abundance of convenience stores. Although there are exceptions, most studies have found convenience stores to be disproportionately located in low income and minority neighborhoods (Walker, Keane, and Burke 2010). In addition, Latinos have been found to buy more food at convenience stores than non-Latinos, and African-Americans are more likely to visit convenience stores than any other ethnic group (Hoffman

2013). The products offered at these stores are typically high in sugar, fat, and salt and linked to diet-related diseases such as diabetes, hypertension, and cardiovascular disease. My colleagues and I found that California schools located near convenience stores were more likely to have higher rates of over-weight students when compared to schools located farther away from such stores (Howard, Fitzpatrick, and Fulfrost 2011). Other studies have reported similar findings (Powell et al. 2007; Williams et al. 2014) or linked residential proximity to convenience stores with higher consumption of sugar, fat, and calories (He et al. 2012; Deierlein et al. 2014).

Concerns about diet and health in disadvantaged populations have spurred advocacy efforts to improve convenience store food offerings. One example is the national Healthy Corner Stores Network, which encourages retailers to sell fresh produce and other more nutritious foods, using methods such as providing grants or loans for refrigerators. Some of these initiatives have struggled to change the culture of convenience stores, as their low-wage employees would prefer not to deal with the complexity of maintaining perishable products, in contrast to the typical shelf-stable offerings (Olender 2007). Others, however, have successfully increased the availability of products such as fresh fruit and low fat milk (Cavanaugh et al. 2014). 7-Eleven has also responded to the growing interest in healthier foods and announced its intention to stock more fresh produce and organic products (Horovitz 2013).

Reducing ownership, increasing control: fast food restaurants

Franchising is even more prevalent in the fast food industry than in the convenience store industry. Nearly all major fast food chains rely on these arrangements and, due to weakened antitrust regulations, many are simultaneously reducing their direct ownership of retail outlets and increasing control of their franchisees. The strategy of offering fewer rewards and placing more risks on franchises has been critical to dominant firms' efforts to increase their power. Some of the anti-competitive strategies used by these firms include: (1) dictating retail prices to franchisees, (2) requiring their franchises to buy ingredients from the franchisor or its approved suppliers, which limits price competition for these products, (3) mandating that if one product from the franchisor is desired it must be purchased in conjunction with additional, unwanted products (Grimes 1999).

The fifth-ranked chain, Burger King, has taken this strategy the furthest, following an ownership change in 2010. The investment firm that acquired the chain that year is led by secretive Brazilian billionaires Jorge Paulo Lemann, Marcel

Herman Telles, and Carlos Alberto Sicupira—they were also the forces behind the global dominance of beer firm AB InBev. Burger King reduced the number of corporate-owned stores from 11 percent to well below 1 percent, while simultaneously requiring franchisees to pay the majority of expenses for remodeling (D. Leonard 2014). The financial success of this strategy led Wall Street analysts to encourage other dominant chains to sell off their corporate-owned stores, which typically accounted for well over 10 percent of outlets (D. Leonard 2014). In 2012, Yum! Brands followed this example, reducing its ownership of stores in

Box 2.2 Adapting to Local Cultures: Yum! Brands Inc.

Fast food has been criticized for weakening local cultures. In France, for example, José Bové famously dismantled a McDonald's that was under construction, criticizing the firm's *malbouffe*, or junk food, and its potential to harm French food habits. Cultural exchange is not as unilinear as some analysts portray, however. Research suggests that although youth do adopt ideas from fast food, their allegiance to their own culture is much stronger (Seubsman et al. 2009; Bugge 2011). Yum! Brands, for example, has had to adapt menus to local tastes, even as local tastes have conformed to the chain's efforts to achieve global uniformity. Their KFC subsidiary in China offers a number of rice dishes and a tree fungus salad, and Chinese Pizza Huts have squid and shrimp toppings (Kaiman 2013). Some analysts have suggested, however, that examples like these are merely "local enactments of globalization" (Tsing 2009, 155); they do little to undermine the underlying logic of "McDonaldization," or capitalist efforts to increase efficiency, calculability, predictability and control (Ritzer 2013).

To compete with the dominance of McDonald's, Yum! Brands has engaged in some significant mergers and acquisitions globally. Its brands include KFC, Pizza Hut, Taco Bell, Pasta Bravo, and Wing Street. The firm has more than 39,000 outlets in over 130 countries. It has been very aggressive in expanding in China, and more than half of the firm's revenue and profits are from these outlets (Baertlein 2013). Yum! Brands had to deal with a number of food scandals in this country, however. In 2005, its Chinese suppliers were found to be using a Sudan Red dye not approved for food use. In 2013, a few months after it was reported that its chicken tested positive for high levels of antibiotics, the firm was forced to drop more than 1000 suppliers in the country. The chain was subsequently reported to have received repackaged, expired meat from a Chinese supplier in 2014 (Associated Press 2014).

Yum! Brands, like many fast food firms, appeals to popular culture in its marketing efforts, such as tie-ins with movies. The 1993 film *Demolition Man* imagined a dystopian future in the year 2032, in which all restaurants, including the very upscale, were owned by a single firm. Incredibly, the chain paid for product placement in order to be named as the monopolist, and even featured the clip, "Now all restaurants are Taco Bell," in its television advertisements. For some non-US markets the reference in the film was changed to Pizza Hut (Mertes 2013). The firm's global domination may not arrive quite so soon, however; although it has more stores worldwide, Yum! Brands remains a distant second in the industry rankings, with a market capitalization that is just one-third the size of McDonald's.

markets where sales were declining, and using the proceeds to invest in ownership of new stores in very fast growing markets (Ashworth 2012), particularly China (Box 2.2).

Although growth in the United States has slowed for the industry as a whole, dominant firms have steadily increased their sales and market share, with the top four controlling nearly 43 percent of the market by 2012 (Table 2.3). The leading firm, McDonald's, has an 18.6 percent market share, but it is close to 50 percent for the more narrowly defined hamburger segment. The overall market is extremely large; 11 percent of calories consumed in the United States from 2007 to 2010 were from fast food, although this is a decline from a peak of 13 percent in the mid-2000s (Fryar and Ervin 2013). Sales totaled $191 billion in 2013 (Statista 2014a).

As with supermarkets, top fast food firms exert incredible power over suppliers, and have helped to concentrate the industries that produce ingredients for

Table 2.3 US fast food restaurant market, 2012

Firm	Market share (percent)
McDonald's	18.6
Yum! Brands (KFC, Pizza Hut, Taco Bell, etc.)	12.6
Doctor's Associates, Inc. (Subway)	6.7
Wendy's	4.8
	CR4: 42.7

Source: IBISWorld (2013).

these chains (Gereffi, Lee, and Christian 2009). McDonald's demand for uniform potatoes in the 1960s, for example, led J.R. Simplot to develop water and energy-intensive irrigation systems rather than utilizing traditional dryland farming techniques. His firm also built a factory specifically to process and freeze potatoes for McDonald's French fries. Simplot has become one of largest private corporations in the United States, and one of just three firms that controlled 70 percent of global potato production by 2003 (Makki and Plummer 2005). The giant meat processor Tyson also developed factories specifically to manufacture McDonald's chicken nuggets beginning in the 1980s (C. Leonard 2014).

Another similarity with supermarkets is conflicts over wage issues. Although the industry has long offered minimum wage jobs and is a model for the deskilling of labor, demands for better pay and working conditions have increased. One key issue is the vast disparity in pay between average workers and CEOs of fast food firms, which rose to a ratio of more than 1,200 to 1 by 2012, making it the most unequal sector of the US economy (Ruetschlin 2014). A number of one-day strikes and protests against poor wages have been organized against leading fast food firms since 2012, seeking more livable hourly rates (Horovitz 2014). McDonald's also received bad publicity when information from its employee hotline was revealed to be both out of touch and hypocritical—the firm provided advice to its minimum wage employees on how much to tip their swimming pool cleaner or nanny, encouraged employees to apply for government food and health assistance, and suggested avoiding fast food due to the risk of becoming overweight (Nicks 2013; Velasco 2013).

The fast food industry also shares key similarities with convenience stores, with a disproportionate number of locations in low income minority areas and near schools. Although public health scientists have only recently documented these disparities (Austin et al. 2005; Fleischhacker et al. 2011), in the 1960s McDonald's Ray Kroc flew in his Cessna aircraft to spot schools so that he could locate restaurants nearby (Schlosser 2001). Studies have linked proximity to fast food outlets to increased consumption of sweetened beverages, as well as higher rates of obesity (Davis and Carpenter 2009; Fleischhacker et al. 2011). Although many chains have attempted to increase the number of healthier offerings on their menus, these tend to be priced higher than their foods high in sugar and fat.

Some of the strongest criticism has been leveled at the leading chains' efforts to market to children. McDonald's spends well over a billion dollars per year on advertising in the United States, with approximately 40 percent of its television placements aimed at children (Bernhardt et al. 2013; Statista 2014b).

Its effectiveness was demonstrated with a magnetic resonance imaging (MRI) study of 10–14 year olds, which found the pleasure center in their brains to be activated by food logos (Bruce et al. 2014). Some local governments have heavily regulated fast food marketing efforts to counter this influence, such as San Francisco's ban on including toys with unhealthy children's meals (Otten et al. 2014).

Although the power of dominant fast food firms has increased in recent decades, this trend has also encouraged alternatives. One of the fastest growing segments of fast food retailing is food carts, which are typically operated by entrepreneurs who lack the capital for even a single brick and mortar restaurant. Although traditional restaurateurs have sought to outlaw food carts with the argument that their fixed costs are lower, giving them an unfair advantage, recent regulatory changes have reduced barriers to entry in many areas (Burningham 2010). Consumers have benefited from the innovative cuisine served at many food carts, such as Korean tacos (essentially Korean food served on corn tortillas), Japanese hot dogs, and maple bacon ice cream. Small-scale farmers may also benefit as some food carts seek to distinguish their cuisine by purchasing high quality, local produce and adjusting their menus to what is in season (Russell 2011).

Revitalizing regulation?

While antitrust regulations have been reinterpreted to give food retailers much more freedom to create oligopolies, there are some limited examples of regulators blocking proposed acquisitions, at least initially. In addition, there have been some investigations into alleged anti-competitive practices, such as fees charged to food manufacturers. Do these signal the possibility that antitrust laws remain a viable tool for reversing trends toward concentration? Or have they become just a weak means to counter the most obvious abuses of power?

In 1999, the Federal Trade Commission (FTC) challenged the Dutch firm Royal Ahold's proposed $1.75 billion acquisition of Pathmark, suggesting the need to sell as many as half of Pathmark's 132 stores in order for the deal to be approved. The rationale for this divestment was that Ahold already owned the Giant and Stop & Shop chains, which overlapped with Pathmark in parts of New York and New Jersey. As a result, Ahold backed out of the deal. A very similar result occurred the following year when the FTC challenged Kroger's proposed acquisition of Winn-Dixie stores in Texas and Oklahoma due to overlapping markets, and the deal fell apart. Both Ahold and Kroger had made significant acquisitions prior to these moves, with little FTC opposition. Ahold spokesperson Hans Gobes suggested that the FTC "changed their model; that's important for the U.S. supermarket sector to realize" (Orgel, Merrefield, and Ghitelman 1999). Many supermarket firms significantly slowed their acquisition efforts after these setbacks, with Kroger, for example, acquiring a 15-unit chain in 2001 but making few other acquisitions until 2014.

After succeeding in dampening acquisitions in the supermarket industry, the FTC held a public workshop on "slotting fees" in 2000. These are fees that many leading supermarket chains require suppliers to pay to gain or maintain access to shelf space, including higher amounts for more prominent locations. Smaller or newer firms may be unable to pay the sums required to obtain access—potentially totaling tens of millions of dollars to place a few products in supermarkets nationally—and may even be shut out completely if competitors pay to exclude similar products. The result is a product selection that is determined by the power of the suppliers, not consumer demands. Due to the secretive nature of these agreements—some of the witnesses at this hearing wore hoods to conceal their identities—the FTC concluded more information was needed. As a result, the agency did not issue specific guidelines, other than identifying several areas that deserved more scrutiny in existing enforcement efforts (Federal Trade Commission 2001).

In 2007, the FTC tried to block another supermarket acquisition. Whole Foods announced it was taking over its main competitor in the natural/organic retail industry, Wild Oats. Whole Foods and Wild Oats had each made more than a dozen acquisitions of competitors in the preceding two decades, leaving no other such chains with a national scope. The FTC had allowed much larger combinations in other industries—Whirlpool merged with Maytag in 2006, for example, resulting in a national market share of 50 percent or more for dishwashers, refrigerators, washer and dryers (Brock 2011). Although such approvals are typically justified by defining markets very broadly, in the Whole Foods-Wild Oats case the FTC defined the market quite narrowly, as "premium natural and organic supermarkets." Whole Foods countered that they were increasingly competing with many other retailers in natural/organic space, including Walmart, which had announced the previous year it was aggressively expanding its organic food offerings.

Wall Street analysts were somewhat puzzled by the FTCs actions but suspected a political motivation. A major investor in Wild Oats was supermarket magnate Ron Burkle, who had become a billionaire by buying and selling supermarket chains. He stood to increase his fortune with the Whole Foods acquisition. Burkle was also a major donor to the Democratic party, as well as a good friend of former president Bill Clinton, while George W. Bush, a Republican, was residing in the White House at the time. Whole Foods fought the injunction, which eventually resulted in the government allowing the deal to go through with some minor concessions, such as selling some stores and giving up the Wild Oats name (MacPherson, Mello, and Rinehard 2009). Wild Oats later became a private label natural/organic brand carried by other supermarkets, including Walmart in 2014.

Another example that raised the hopes of antitrust advocates was a series of five workshops held jointly by the USDA and DOJ in 2010, which addressed enforcement issues in a number of stages of the food system, including retailing. The report of these workshops acknowledged the many abuses of power used by large firms, such as bid rigging, market manipulation, and one-sided contracts. Ultimately, however, its authors suggested that due to the way laws were interpreted by judges, there was little possibility of using antitrust enforcement to address these issues (Department of Justice 2012).

Together, these cases demonstrate that government is capable of blocking the expansion of power of dominant firms, but these are relatively rare exceptions. Such power has been infrequently exercised in recent decades, even for cases with strong evidence of anti-competitive behavior; a significant strengthening of antitrust enforcement therefore appears to be unlikely. This is not surprising when taking into account research on the policy process in the US, which indicates that the textbook version of "democracy" applies quite well to powerful business interests, but average citizens and broad-based interest groups, in contrast, have little to no influence. While there are exceptions, particularly when the legitimacy of a government is threatened, the majority of policy actions are limited to the consensus of economic elites (Bartels 2010; Schlozman, Verba, and Brady 2012; Gilens 2014).

This chapter described how retailers have responded to the weakening of US antitrust laws since the 1970s. Most firms have exploited the freedom to exact steep discounts from suppliers, particularly Walmart, while others have also focused on the increased latitude to acquire competitors. In the convenience and retail sectors, a greater ability to exploit franchisees is an additional strategy that has been used to increase power. The position of retailers as the closest link to consumers helps provide greater leverage over upstream segments of the food system—which would otherwise find it difficult to connect to them—such as charging slotting fees. The following chapter explores how distributors are also using their structural position to maintain or even increase their dominance.

Chapter 3

Structuring dependency: distribution

*Eliminating the middleman is never as simple as it sounds.
'Bout 50% of the human race is middlemen, and they don't
take kindly to being eliminated.*
—Malcolm Reynolds (*Firefly*)

Distribution is a part of the food system that tends to be quite hidden, unless you pay attention to the names on the trucks in the alleys behind supermarkets and restaurants. It is also more competitive than retailing or processing, the two segments of the value chain that it connects. Of the estimated 15,000 plus distributors in the United States, for example, most have less than 100 employees (Blissett, Kahn, and Boyce 2008). As with packaged organic foods discussed in Chapter 8, venture capitalists see an opportunity in a market they consider fragmented and are acquiring firms with a goal of consolidating food distribution. The largest wholesaling and distribution firms are increasing in size as a result of this economic concentration, but they are also facing potential threats to continued growth. Brick and mortar retailers, as well as web-based retailers, are eroding their markets through self-distribution or receiving products directly from producers. This transformation, called *disintermediation*, is affecting both broadline firms, which supply nearly everything to retailers, and specialty firms, which distribute particular foods or beverages, such as alcohol, seafood, dairy, meat, produce, etc.

How have dominant distributors managed to respond to this trend, and even continued to increase their power? A *network exchange* perspective helps explain why their strategies have been successful. By analyzing the connections between people or firms, network mapping can show the sometimes hidden structure that influences their interactions. Harvey James, Mary Hendrickson, and I (2013) explored several types of food systems with this approach. Rather than the broad concept of power, we focused on *differential dependency*, which is more measurable at the micro-level. It refers to relationships that are unequal, with one party more dependent on the other. When characterizing the network, it is also important to understand which connections are *contingent*, meaning that engaging in an exchange also requires that there be a complementary exchange with another part of the system. From a retailer's perspective, for example, an exchange with a distributor will typically require that distributor to also engage with an upstream food manufacturer, processor or farmer. The dependency of a given node can therefore be affected by exchanges that occur many steps removed from direct connections.

This chapter explores how distributors in the broadline and several special- ist segments attempt to structure their networks to maintain their relevance, as well as the resulting impacts on suppliers and customers. Broadline firms are by far the largest in the industry and control approximately 60 percent of sales (McConnell 2014a). Beer is another focus, due to an extra layer of government regulation in the United States, which helps shape a very unique distribution system compared to most other specialists. A third segment of interest is the tiny but rapidly growing "values-based value chains"—these distributors, which include those working with "food hubs," place more emphasis on non-economic goals, such as ecologically sustainable production practices. As a result, they may be more amenable to reducing the differential dependency between farm- ers and retailers rather than increasing it.

Networks, power, and distributors

Distributors are a classic example of a "middleman"—these types of nodes can enable exchange but can also block it if two segments of a system are unable to connect without them. A retailer in New York City that wants to sell a tropical fruit, such as bananas, will typically be unable to connect directly with a banana plantation and must buy from a distributor. If there are numerous distributors to choose from, the retailer may be able to engage with them on relatively equal terms—this is the basis for the definition of competitive markets discussed in Chapter 1—but not if there is only one. What this definition does not fully incorporate, however, is that the types of exchanges between suppliers and distributors also cascade through the network to affect retailers and other parts of the network (James, Hendrickson, and Howard 2013). If two banana producers merge, for example, the potential impacts of higher prices resulting from reduced competition could be passed from distributors to retailers. Modifications to other contingent exchanges incorporated into foods may have similar impacts, such as rising prices for fossil fuel required for distributors' trucks or the sale of herbicides tied to seeds purchased by farmers (see Chapter 7).

One way to increase the dependency of retailers on distributors is to become the only provider of a desirable product. If retailers want access to this type of product, such as a particular brand of imported wine, they have to deal with the exclusive distributor, who monopolizes the contingent exchange with the producer. When contingent exchanges are not exclusive, there is more competition, and as a result, the retailers experience less dependency. Their dependency can be reduced further if the retailer removes the potential barrier of a distributor and makes a direct connection with the supplier. However, this is likely to result in the additional costs of logistics typically handled by distributors, such as finding consistent suppliers, negotiating exchanges, storing (perishable) products, transportation, and delivery.

Walmart is an example of a retailer that has taken this approach at a very large scale. The corporation has vertically integrated and created an enormous self-distribution system, shrinking many previous distribution channels as it has increased its market share in recent decades. Walmart was the first retailer to introduce cross-docking, which significantly reduces the amount of warehouse space that is otherwise required (Gereffi and Christian 2009; Baines 2014a). It involves each producer delivering to one side of a distribution center on a precisely timed schedule. The products are immediately moved to the opposite side of the facility and commingled with other goods to fully load trucks destined for retailer stores. Savings from this model helped determine the growth of new Walmart store locations, which were typically within a very tight radius of distribution centers (Holmes 2011). By 1999, forty-seven of the fifty largest retailers had adopted cross-docking (Martinez 2002). More recent entrants into food retailing have also followed suit, including Target, which opened its own food distribution centers in 2008 (Blissett, Kahn, and Boyce 2008).

Efforts to cut out the middlemen do not always go smoothly, as Malcolm Reynolds noted in *Firefly*. When Quaker Oats bought the beverage company Snapple, for example, the firm was already successfully delivering its previously acquired brand, Gatorade, directly to supermarkets. Executives believed they could bring Snapple products into this system. Quaker tried to convince the numerous independent Snapple distributors, who supplied primarily convenience stores and gas stations, to hand over their high profit margin Snapple rights with supermarkets. In return, they would receive the right to distribute low profit margin Gatorade to their smaller retailers—the bargain was even more unfair because Gatorade did not sell as well as outside of supermarkets. Not surprisingly, this strategy met with little success (Gadiesh et al. 2001). The distributors, who had played a large role in building the Snapple brand, lost their enthusiasm for the product by the time Quaker reversed its emphasis on self-distribution. Just over two years after acquiring Snapple, Quaker was faced with steeply declining sales—although advertising missteps also contributed—and sold the division at a $1.4 billion loss. After this sale to a holding company for $300 million, the brand was acquired by Cadbury Schweppes and later spun off as part of the Dr. Pepper Snapple Group. The Quaker-Snapple case is now widely used in business schools as an example of what not to do when acquiring other firms (Deighton 1999) but could also be viewed as an example of how a distributors' strategic location within the structure of a network can allow it to challenge the power of a much larger manufacturer.

Successful challenges are relatively rare, however, and most distributors, aside from the very largest, are experiencing declining sales. Dominant firms are achieving growth, but primarily as a result of acquisitions. Mid-size firms in the industry are experiencing the brunt of the impacts of disintermediation, and a significant number are likely to: (1) be acquired by dominant firms, (2) become more specialized, or (3) go out of business.

Buying out the competition: broadliners

Broadline distributors are also called "full-line" distributors because they carry a very wide variety of products. By engaging in so many contingent exchanges, they provide customers the convenience of potentially relying entirely on just one firm rather than dealing with separate distributors for meat, produce, dairy, and numerous other products. Some 7-Eleven stores may have as many as fifty or sixty deliveries a week, for example (Wulfraat 2014). The convenience of utilizing a single distributor may require tradeoffs, such as more limited selections in product categories. Perhaps more importantly, there is less potential for retailers to pit distributors against each other in order to receive better contract terms; many restaurants in the United States, for example, engaged in this strategy with the largest broadline distributors, such as Sysco and US Foods, to negotiate lower prices or improved service (Bartz 2014).

In December 2013, however, these two firms announced their intention to combine, with Sysco acquiring US Foods at a price of $8.2 billion in stock, cash and assumption of debt. The majority of funding was loaned by the Wall Street firm Goldman Sachs, which also funded the merger of nine distributors to form Sysco in 1969. The US Foods buyout was strongly opposed by public interest groups, such as Food & Water Watch, but most analysts expected it to be approved by the Federal Trade Commission, perhaps with some divestitures as conditions (McConnell 2014a). The agency surprisingly filed an injunction to prevent the merger in early 2015 but also indicated that if the judge does not approve this request it may drop its challenge altogether (Bartz 2015). If allowed, the combination would increase Sysco's share of the broadline market from 24 percent to approximately 36 percent (Table 3.1). Sysco representatives claimed they would continue to face thousands of remaining competitors in the distribution industry, but with the acquisition of US Foods, the next largest would be less than one-sixth the size of Sysco.

From a network exchange perspective, Sysco is planning to remove the only similar node, that is, the one other broadline distributor with a fully national scope and offering a selection of hundreds of thousands of items. For customers that rely on this type of exchange, which range from institutional cafeterias to expensive gourmet restaurants, they will have only one option available instead of two. Not all retailers are this dependent on the biggest firms, however, and have demonstrated their willingness to give up such convenient access to a vast product selection if sufficiently motivated, as a controversy in California demonstrated.

In 2013, a San Francisco Bay Area television news investigation revealed that Sysco was using outdoor, unrefrigerated metal sheds to drop off perishable foods, including raw meats, milk and vegetables, at fourteen sites in the area. Hidden cameras showed that the food remained in these sheds for up to five hours (Nguyen et al. 2013). The company admitted that it had kept the practice hidden from state regulators for years, and the California Department of Public Health subsequently uncovered additional sheds, including at other Sysco distribution centers. Food safety scientists suggest foodborne illness risks were greatly increased by these practices, and the incident triggered increased sales for Sysco's smaller competitors (McConnell 2014b).

Unfortunately for retailers, some of these competitors may end up being taken over by Sysco, which has acquired nearly 150 other companies or divisions

Table 3.1 US broadline distributor market, 2013

Firm	Sales in billions	Market share (percent)
Sysco	$44	23.9
US Foods (acquired by Sysco, 2013)	$22	11.9
Performance Food Group	$13	6.9
Gordon Food Service	$10	5.2
		CR4: 47.9

Source: Hauter (2014).

since its founding (Carnevale 2014). Some of the more recent transactions are shown in Figure 3.1. In industries with declining sales and profits, acquiring other firms is a key strategy for increasing power. It has contributed to Sysco growing large enough to be a component of the S&P 500, with a market capitalization of approximately $21 billion in early 2014, and a ranking of 204 on the Fortune 500 list of the world's largest companies (Cacace 2014). In early 2013,

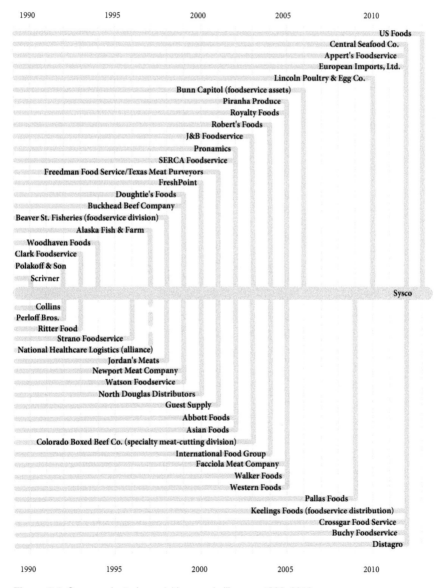

Figure 3.1 Sysco: selected acquisitions and alliances, 1990–2013.

Sysco's CEO expressed a strong interest in buying Performance Food Group if the firm was for sale (Collings 2013). Whether or not the acquisition of US Foods is allowed by regulators, there is little reason to think Sysco would be discouraged from continuing to buy other competitors. Second-tier firms are also making acquisitions, such as Gordon Food Service's 2014 buyout of Glazier to increase its geographic scope, and a handful of takeovers by Performance Food Group (see Box 3.1). Such activity by the largest firms has contributed

Box 3.1 Catering to the Big: Performance Food Group

Consolidation is accelerated when suppliers or customers in upstream or downstream industries become much larger, as this pressures firms to achieve enough size to have negotiating leverage with these connections. In other words, "only the big can serve the big" (Hannaford 2007, 30). Performance Food Group (PFG) is a good example—the firm started in 1875 as a distributor of canned fruits and vegetables to restaurants and grocery stores, and has grown into the twenty-second largest private company in the United States (Forbes 2014). PFG currently supplies national chain restaurants, such as Cracker Barrel, Ruby Tuesday, Outback Steakhouse and T.G.I. Friday's, which account for a large percentage of the firm's revenues. Although these customers have the power to demand much lower price markups than PFG's smaller customers, they also receive additional services, such as priority distribution routes. Smaller customers, such as independent restaurants, are likely to pay higher margins and receive less preferential service.

PFG has accelerated its growth through (1) an initial public offering, (2) selling a food processing division, and (3) accepting investments from venture capitalists to become private and then using the proceeds of these transactions to acquire a number of smaller competitors. The firm became publicly traded on the NASDAQ exchange in 1993. In 2005, PFG sold its highly successful packaged salad division, Fresh Express, to the global produce giant Chiquita, for $855 million, in order to focus entirely on distribution. Three years later, PFG was acquired by a pair of private equity firms, The Blackstone Group and Wellspring Capital Management, for $1.3 billion (Blissett, Kahn, and Boyce 2008). Food distributors that had already been acquired by these venture capitalists, Vistar and Roma Foods, were then folded into PFG. Recent acquisitions include Somerset Food Service in 2009; Ledyard in 2010; and Institution Food House, Fox River Foods, and Liberty in 2012. The firm now has twenty distribution warehouses and operates nationally but remains much smaller than Sysco and US Foods (Performance Food Group 2012).

substantially to raising the CR4 for broadline distributors more than 10 percent in the decade from 2003 to 2013 (Hauter 2014).

Increasing scale can be an advantage in an industry with expensive infrastructure requirements and low profit margins. The differential dependency of suppliers compared to very large distributors may result in negotiating lower prices and subsequently enables these distributors to undercut the prices of their competitors. This is particularly evident for those that have developed private label products as alternatives to more expensive name brands. Sysco, for example, offers several of their own brands for meat and ethnic food categories, as well as brands for imported foods, dairy products, coffee, and bread. This power can potentially result in a "waterbed" effect if the low prices negotiated by large distributors encourage suppliers to make up for these lost profits by raising their prices for other distributors—these disparities could be continually magnified as more and more retailers defect to the largest firms as a result (James, Hendrickson, and Howard 2013).

One response from a group of smaller, regional distributors has been to form a cooperative, called Distribution Market Advantage (DMA). By pooling some resources, such as purchasing power and geographic coverage, but remaining independent firms, DMA is growing faster than Sysco or US Foods; the group serves an increasing number of national restaurant chains, for example (McConnell 2014b). Distribution cooperatives are more common among grocery retailers, and account for four of the ten largest cooperatives in the United States (Snyder 2011). The largest of these, Topco, is vertically integrated to produce and distribute foods to more than fifty members, which include Wegman's and Meijer. Topco offers more than twenty different private label brands, such as Full Circle for natural and organic products. Because most participating retailers are in non-overlapping regions, consumers may perceive these brands as exclusively owned by their local chain (Howard 2009b).

Protecting legal monopolies: beer

Beer distributors have been much more protected from disintermediation than other specialists, due to the three-tiered system of alcohol distribution enacted after the repeal of Prohibition in the United States in 1933. It separated producers, distributors, and retailers, so that producers could not sell directly to retailers. These regulations were designed to prevent abuses that occurred before Prohibition, when bars accepted loans or supplies in exchange for product exclusivity agreements. These agreements frequently pressured retailers to increase beer sales through customer overindulgence and were a key factor in the growth of some brewers, such as Anheuser-Busch, at the expense of competitors (Knoedelseder 2012).

Although retailers are no longer tied directly to producers, most of the approximately 3,000 beer distributors have exclusive territory agreements with brewers (Crowell 2013), which can be viewed as local monopolies on these contingent exchanges. As the beer industry consolidated (see Chapter 4), the distributors that were linked to the surviving producers increased in importance, and retailers selling beer were left with fewer sourcing options. This results in differential dependency for small retailers, who cannot afford to risk losing access to popular beer brands, particularly those owned by the two largest producers in the United States, Anheuser-Busch InBev and MillerCoors. Distributors, on the other hand, face little risk if they lose a retailer that accounts for just a tiny portion of their business.

Exclusive territories also create bottlenecks for many small brewers, who are either unable to access key markets, or must sign contracts that are overwhelmingly favorable to the interests of the distributors. These firms have been known to abuse this power, such as preventing craft brewers from exercising contractual rights to leave. Such actions have resulted in court battles that cost breweries, such as Brooklyn Brewing and Dogfish Head, hundreds of thousands of dollars in legal fees (Hindy 2014). Bell's Brewery in Michigan left the lucrative, nearby Chicago market for several years after their distributor planned to sell the distribution rights to another firm, which did not require approval from the brewer. Although illegal, it is reportedly a common practice among distributors in Chicago to provide free products for retailers, a subsidy that makes it difficult for small brewers to gain access (Day 2006; Ylisela, Sterrett, and MacArthur 2010).

The three-tier legal barrier to disintermediation is beginning to break down, however, as a result of pressure from dominant retailers and brewers, particularly Costco and AB InBev. Twenty states now allow brewers to self-distribute, and AB InBev has capitalized on these changes in recent years by acquiring more distributors. The brewer has quickly become the largest distributor in the country (Table 3.2), with an estimated market share of 8 percent in 2012, and some analysts predict their share could rise above 25 percent (Logan 2012). In states where self-distribution is not allowed, the corporation is exercising more indirect control by encouraging distributors to consolidate, and to align with AB InBev's goals (see Box 3.2), such as giving up the distribution of competing craft beers (New America Foundation 2012). Approximately 60 percent of AB InBev's distributors focus exclusively on the firm's products (Blanding 2011).

Table 3.2 Leading US beer distributors, 2013

Firm	Sales in billions
AB InBev	$3
Reyes Beverage Group	$2.2
Silver Eagle Distributors	$1
Ben E. Keith Beverages	$0.8

Source: Schumacher (2014).

Box 3.2 Targeting Ethnic Groups: Silver Eagle

Silver Eagle is an AB InBev distributor located in Texas. The firm uses extensive consumer demographic data to target markets based on ethnicity, such as distinguishing immigrants from northern or southern Mexico, or Central or South America, particularly for their portfolios of imported Mexican and Asian beers (Hildebrandt 2012). The data mining system, called BudNet, was developed by Anheuser-Busch—it requires participation from distributors and retailers to increase sales for all three segments of the supply chain (Kelleher 2004). This system not only collects sales data, but on a daily basis it analyzes information such as shelf space, the location and visibility of displays, the price at neighboring stores, and census data on neighborhood income and ethnic composition. Marketing strategies are continually fine-tuned at very small geographic scales, such as stocking more single-serve packaging in locations where these are expected to increase purchases or using attractive models on billboards to match a community's predominant ethnicity.

The practice of focusing on ethnic minorities, particularly marketing malt liquor to African–American communities, has been criticized for encouraging underage drinking and contributing to the negative health consequences of overconsumption (Hackbarth, Silvestri, and Cosper 1995; Alaniz and Wilkes 1998). Latinos are particularly sought after in marketing efforts because their population is growing (Kelleher 2004). Silver Eagle's location in Texas, where immigration rates are higher than other states, has contributed to its growth in an industry that is experiencing flattening sales nationally. Silver Eagle has also increased its size through key acquisitions, including BudCo in 2007, and distribution rights for a number of craft and import beers from Glazer's in 2009. AB InBev brands make up more than 85 percent of their product mix, however (Hildebrandt 2012).

As noted above, there are approximately 3,000 independent beer distributors in the United States, but perhaps less than 1,000 are of significant size (Hindy 2014). In addition, there has been a loss of approximately 1,600 distributors since the early 1990s, despite a rapid increase in the number of breweries during the same time (Furnari 2014). Although distributors frequently exercise more power in exchanges with small retailers, they have not consolidated enough to address their own power imbalances when negotiating with dominant brewers. Due to the very one-sided contracts required to obtain exclusive rights from these corporations, they cannot switch to another product nor even sell their own business without brewer approval. The New America Foundation (2012) notes that distributors are increasingly forced into a tournament type of system, where their performance is compared to other distributors; the "winners" are rewarded and the "losers" are punished, potentially even losing their contracts.

Although concentration among brewers has had negative impacts on beer distributors, they continue to exercise tremendous power relative to their size. In 2012, their trade association, the National Beer Wholesalers Association (NBWA) was the third largest political action committee in the United States (Ascher 2012). The group lobbies extensively to maintain the three-tier system and the near monopolies it provides them. The rhetoric employed to support this position is that regional distributors are needed to: (1) provide jobs, (2) navigate very different state alcohol laws, (3) maintain a level playing field for smaller brewers to obtain retail access, and (4) accurately collect taxes for the government (NBWA 2014). Because the system was designed to increase the price of beer in order to discourage consumption, it remains a very lucrative segment of the beer value chain. Cindy McCain, for example, the wife of senator and former US presidential candidate John McCain, inherited absentee ownership of Hensley & Co., one of the largest AB InBev distributors in the country. Her wealth was estimated to be $100 million in 2008 (Associated Press 2008).

Distributors have had mixed success in maintaining their legal monopolies, however. In addition to losing battles over legalizing self-distribution in some states, the majority of states allow exceptions to the three-tier system for small brewers. This allows an increasing number of microbreweries to sell their beer directly, without giving a cut to the middleman. In Florida, this led a lobbyist for Anheuser-Busch InBev distributors to promote a bill that would require microbreweries to pay distributors a 30–40 percent markup on bottles and cans before selling them, even if they never left the brewery. He found a legislator who had accepted campaign donations from three of these distributors and was willing to sponsor the bill. The attempt failed, however, after attracting negative attention nationally. Distributor's lobbyists in the state had already succeeded in banning to-go sales in half gallon containers, which is the most popular size of "growler" in the United States, although both larger and smaller sizes were allowed (Liston 2014).

Recent legal changes have created new niches in the beer distribution business, including for larger craft brewers to distribute competitors' products. Craig Purser, president and CEO of the NBWA said that although the distribution industry is consolidating, "what we're seeing now is a number of new (craft or specialty) distributors coming to market. The cost of entry to distribution is very low" (Crowell 2013). Stone Brewing, for example, is based in San Diego and is ranked the tenth largest craft brewer in the United States. The firm self-distributes its own beers throughout Southern California and also carries thirty-five additional craft brews while turning down requests from many more. Stone recently partnered with Maui Brewing to distribute all of their mainland brands in Hawaii (Peña 2013). Not all craft brewers have found success in the distribution channel. Brooklyn, for example, lost too much money and sold its distribution business in 2003 (Hindy and Potter 2011).

Values-based value chains?

Dominant distributors are increasing in size and focusing on their largest suppliers and customers. Not surprisingly, this leaves an increasing number of smaller niches with unmet needs. One of these is the growing movement for a food system that incorporates ecological, social, and other values. Although organic and fair trade products can now be found in nearly any mainstream outlet, foods and beverages embodying other such criteria are more difficult to find. Some initiatives have therefore recognized the critical need to create a distribution infrastructure, so that producers and consumers with shared values may engage in more contingent exchanges—while Community Supported Agriculture, farm stands, farmers markets, and other direct markets are growing rapidly, they are not of sufficient scale to reach the majority of consumers.

Interest in the criterion of "local," for example, underlies many of these efforts, despite the lack of a clear definition for this term. Some efforts also include other complementary values, such as pastured/grass-fed livestock production, restricted use of antibiotics or pesticides, and a smaller scale of production. The widespread support for local is based on a diverse range of motivations that include taste, freshness, preserving local farmland, supporting local communities, desire to reduce fossil fuel use, and knowing the exact origin of the food (Howard and Allen 2010). Many of these values could be encompassed by the term "quality," which contrasts with the strong emphasis on quantity and low prices in the conventional food system (Goodman 2003).

Despite many shared characteristics, what I refer to as values-based value chains may utilize a number of different names, including food hubs, community food enterprises, short food supply chains, or local supply chains. A significant number are not yet financially self-sustaining and are dependent upon grant funding, volunteer labor, below usual cost access to infrastructure or other means of reducing costs. The emphasis on values, however, has attracted ideal-istic individuals and organizations willing to make these contributions in order to support the creation of new alternatives. Importantly, this includes the consum-ers who are willing to pay more than the typical price for a product.

Do these efforts have the potential to decrease rather than increase the differ-ential dependency of the retailers and suppliers they connect to each other? For some, the answer appears to be yes. An economic development effort in Athens, Ohio, for example, employed a social network perspective to identify many unconnected local food efforts and bring them into the hub of a kitchen incubator—a shared facility for processing and packaging foods. The next step was to "weave" together more social connections between the people involved and to increase the quality of these connections. The result was a more densely connected network and a reduced dependence on the central hub that first brought them together (Krebs and Holley 2006). The organization attributes the much lower failure rate of participants, when compared to similar efforts, to its emphasis on creating a robust, multi-hub network (Ackerman-Leist 2013).

Farm to Institution (FTI) efforts are a specific type of values-based value chain that also appear to be successful in reducing dependency in value chains. These efforts connect producers with schools, hospitals and other large foodservice organizations. Large institutions have a high degree of purchasing power, which has typically aligned them with the biggest distributors, but some have revised their low bid contract policies in order to support those excluded from the conventional model. One example is the Agriculture and Land-Based Training Association (ALBA), located in Salinas, California. This non-profit oper-ates a Spanish-language organic farming training program, and its graduates include many former farmworkers. To help ensure markets for these new farm-ers, ALBA formed a cooperative to pool organic produce and distribute it to nearby institutions. Stanford University was among the first customers, but many other universities, hospitals, and corporate cafeterias have since joined this network (Izzo 2007).

For initiatives that are not organized as non-profits or cooperatives, the evidence for reducing dependency is less conclusive. Cherry Capital Foods of

Michigan is one example. The firm, which is located in the cities of Traverse City and Lansing, began in 2007 with just one van. Its focus on helping farmers within the state to find new markets and comply with regulations enabled it to grow rapidly, and by 2014 it operated six refrigerated trucks. The firm links more than 200 producers with grocers, restaurants, and school food service companies, but this success has also enabled the acquisition of several small, specialty distributors in Michigan (Geiger 2014). Cherry Capital Foods is therefore reducing differential dependency for farmers by enabling new connections but also potentially increasing it by removing competing nodes in these networks.

In addition, giant distributors and self-distributing retailers have observed increased interest in local foods, and are attempting to capitalize on this trend. Sysco, for example, markets its dedication to sourcing local produce through a Buy Local, Sell Fresh program. Interviews with farmers prominently featured in these marketing efforts have found that some are not even aware that they are supplying the firm, because their produce first passes through other brokers and aggregators (Falat 2011). This suggests that many of the values implied by Sysco in their efforts to increase retail sales are not well supported at the producer stage of the network. Walmart is also promoting its efforts to sell local produce—which they define as grown and sold within the same state—and claiming that it assists farmers while also saving consumers money. The corporation reported that 11 percent of its produce nationally met these criteria in 2012. Critics, however, have faulted the lack of transparency needed to evaluate the supposed benefits, such as how much consists of commodity crops in very large states like California (Swanson 2014). Skeptics also suggest that Walmart is not fundamentally changing its business practice but conventionalizing and co-opting the movements that advocate local food on the basis of deeper values (DeLind 2011).

This chapter has explored how rapidly growing smaller distributors that focus on higher-value, "quality" foods and beverages, such as local produce and craft beer, are reducing differential dependency for consumers, retailers, and producers—albeit a very small proportion of them. Although many rely on social or economic subsidies, their direct impacts are perhaps not as important as simply demonstrating the potential to oppose the dominant trend of increasing power of larger distributors, like Sysco and AB InBev's distribution businesses. The next chapter explores how packaged food and beverage firms also increase differential dependency for consumers, by decreasing their self-reliance and reducing the number of competitors found in retail environments.

Chapter 4

Engineering consumption: packaged foods and beverages

John Anderton, you look like you could use a Guinness!
—electronic billboard (Minority Report)

Food and beverage manufacturers constantly run up against the limited size of our stomachs in efforts to increase their power. There is only so much we can physically eat or drink and only so much that firms can do to reduce their costs as well. Generating profits that outperform the average for other industries therefore requires continually steering our purchases in new directions. This chapter focuses on two key strategies packaged food firms use to engineer more profitable consumption patterns and increase their market shares, *deskilling* and *spatial colonization* (Jaffe and Gertler 2006; Winson 2013). The most dominant firms and their impacts on society, particularly for public health and consumer prices, are explored in more depth for beer, soymilk and bagged salads. The possibilities for consumers to resist these strategies are also discussed.

Deskilling is a term that is frequently used in reference to labor, when capitalists reshape jobs in order to decrease wages (Braverman 1998). It can also apply when capitalists reshape socio-cultural practices to increase purchases, moving us away from self-provisioning to become mere "consumers" (Jaffe and Gertler 2006). Deskilling increases control for capitalists but makes us more dependent upon them by eroding our knowledge and abilities. Food preparation and cooking practices that were once common have become less prevalent, often through the active efforts of food manufacturers to make buying higher profit-margin processed foods a regular habit. In the early 1900s, for example, home economics courses in schools were heavily funded by food processors, and promoted the use of canned foods instead of fresh foods (Levenstein 2003). These courses also encouraged households to use the same efficiency principles that were deskilling factory workers to save time spent on chores; increased sales of new technologies, such as refrigerators and dishwashers was one desired outcome (Vileisis 2008). Manufactured infant formula was popularized beginning in the 1930s, and the more beneficial practice of breastfeeding was stigmatized. This resulted in a steep drop in breastfeeding rates, particularly for low income and minority women, aided by a US government program that provided free formula (Jaffe and Gertler 2006). Even the knowledge of how to boil water now appears to be at risk, with frozen, microwavable oatmeal being offered by a number of food makers (Hu 2013).

The strategy of spatial colonization occurs in numerous areas, as packaged food and beverage firms extend beyond previous limits and increase the scope of their power (Winson 2013). One obvious form of expansion is to enter new markets, such as less industrialized countries. These populations may have less purchasing capacity than more industrialized countries but are growing faster, and offer new territories for household deskilling. There are additional boundaries that firms push against, however, which include shelf space and mind space (Hannaford 2007).

Although the premises of supermarkets and other types of food retailers have been getting larger, shelf space is still finite. The increasing concentration and power of the retail industry poses a challenge for packaged foods makers who want more of their products on these shelves, and at more desirable locations within the store. The largest firms have a greater ability to pay the "shelf slotting" fees demanded by US retailers, which may total as much as $9 billion annually (Dickerson 2004), and can dominate the prime eye-level shelves and ends of aisles. These manufacturers have resisted offering real diversity to their customers, but they have provided an appearance of diversity, or what business analyst Steve Hannaford (2007, 12) calls "pseudovariety." By offering very slight variations on existing products, they are able to take up even larger amounts of shelf space, crowd out smaller competitors, and prevent new competitors from breaking into key retail outlets. Honest Tea founder Seth Goldman used to mock this product proliferation in the soft drink industry, which "seemed to be variations on vanilla cola, diet vanilla cola, vanilla cherry cola, diet vanilla cherry cola...," before his firm was acquired by Coca-Cola (Goldman 2010).

Mind space, or brand recognition and loyalty from consumers, is another limited resource but one that can be overcome by massive and constant advertising (Hannaford 2007). By commanding larger amounts of our memory, leading brands are able not only to increase sales but also maintain some leverage over retailers. Walmart could not afford to simply drop Coca-Cola or Nestlé brands, for example, if their customers continued to demand these products. As a result, even if we don't consume them, most of us are exposed to numerous advertisements for packaged foods, starting at a very young age. Children view an average of more than twenty-one television ads per hour, and approximately half of these are for food (Taras and Gage 1995). Of these, approximately 88 percent are for foods high in salt, sugar and fat, but these ads are still allowed to be tax-deductible expenses for these firms (Blumenthal 2014).

We are not to the point that billboards are calling us out by name, as in the movie *Minority Report*, but new technologies are bringing us much closer. In 2013, Mondelēz (formerly Kraft), the maker of gum and chocolate candies, began testing sensors on supermarket shelves that determine a customer's gender and approximate age, and then show video advertisements targeted to their demographic (*CSP Daily News* 2013). The sensors can also detect if a product has been removed from a shelf. The data will be relayed to the company for analysis, with a goal of enhancing product placement and in-store promotions. The technology is expected to become public in 2015, and similar devices are also being tested by the UK supermarket chain, Tesco (Peterson 2013).

Brands have colonized our minds to the point that they are seen as more valuable than physical assets. Coca-Cola's brand value is estimated to be worth more than $79 billion, with more than 90 percent awareness in many countries (Interbrand 2013). Some companies are taking this idea to its logical end and becoming virtual food companies. Rather than manufacturing and packaging foods themselves, they outsource these less profitable tasks to other firms, and (like the shoe company Nike) focus only on branding and marketing. This approach carries the risk of making firms even more vulnerable to public relations crises, however, as a loss in consumer confidence could wipe out the brand's value almost overnight (Larkin 2003).

These deskilling and expansion strategies have been effective in increasing the dominance of the largest firms. By 2011, the top ten firms had captured 31.9 percent of all food and beverage sales in the United States, and globally the top ten controlled 15.2 percent (Alexander, Yach, and Mensah 2011). Concentration ratios are frequently much higher for specific products. Breakfast cereal, snack chip, canned soup and soft drink industries, for example, are all dominated by a small number of firms, which typically leads to price signaling and higher profits. The high prices charged by leading cereal firms were criticized in a 1995 report by Samuel Gjedenson, a member of Congress from Connecticut, and Charles Schumer, a senator from New York (Burros 1995). In response to the negative publicity that followed, Post admitted its cereals were priced unreasonably and reduced them by 20 percent, leading to price drops from other leading firms (Cotterill 1999). More recently, cereal, chips and soft drinks have been reported to have profit margins of 70 percent or more above costs, compared to less than 30 percent for most other foods (Warner 2013, 195).

The foods and beverages with such high profit margins are typically composed of heavily subsidized commodity ingredients, such as corn, soy, wheat and dairy. These are then transformed via industrial processes into intermediate food substances, such as high fructose corn syrup and hydrogenated soybean oil, which are used in food manufacturing (Moodie et al. 2013). These intermediate ingredients are nearly impossible to make in your own kitchen but contribute to making processed foods more "crave-able" (Moss 2013). Most of us would call these products addictive, as they are deliberately engineered to have tastes, textures, and synthetic flavors that make them difficult to consume in moderation. This includes mathematically modeling the optimal level of sugar to achieve the "bliss point," and expensive machines that determine the perfect breaking pressure for crunchy foods (ibid.). Interestingly, many of the employees involved in creating these types of products say they do not themselves consume them (Moss 2013; Warner 2013).

In a vicious circle, these ultra-processed products are the focus of advertising efforts and placed in prominent locations in the store, which encourages us to eat more and reinforce these efforts. Such strategies are targeted to specific demographics, such as low income households, minorities, women, and children. The boxes of cereal marketed to children for example, are designed so that the eyes of characters appear to gaze downward at them (Franco 2014). African-Americans are targeted with more processed food advertisements than Americans of European descent; this may contribute to dietary differences, as well as higher rates of obesity and chronic diseases, such as diabetes, heart disease, and stroke, when compared to other ethnic groups (Neff et al. 2009).

Although packaged food and beverage industries offer numerous products, I focus more specifically on just three: beer, soy milk, and bagged salad. Beer firms have successfully engineered consumption increases for decades but are now emphasizing global expansion in response to slowing sales in industrialized countries. Soy milk and bagged salads are more recent innovations tied to the growth of organic foods. Although the leading brands quickly became dominant by monopolizing shelf space in conventional channels, they are now losing market share to competitors and private label products. As a result they have become acquisition targets for even larger firms, which illustrates trends seen in numerous other packaged food and beverage categories. These trends have been resisted by a segment of consumers, however, who are challenging the negative impacts of concentration and attempting to become less dependent on the largest firms.

A diverging market: beer

In the late 1800s beer was a very local product. It presented a number of barriers to concentration, including being expensive to store and transport due to its weight, perishability, and a simple composition—water, malted grains, hops and yeast—that could be produced at a very small scale. At that time, there were well over two thousand breweries in the United States, despite a much lower population. By 1979, however, there were just forty-eight brewing facilities in the entire country (Shin 2011). Prohibition in the 1920s played a key role, by putting all breweries out of business for more than a decade. Larger firms that had diversified into other products, such as ice cream, soft drinks and candy, were better positioned to re-emerge when prohibition was repealed (Van Munching 1997).

In addition, brewing your own beer at home was illegal in every state from 1933 until 1979. This government-enforced deskilling became a significant barrier to entry, which allowed the remaining firms to become increasingly less innovative. Until recently, nearly all of the large brewers were producing virtually the same product—a pale US-style lager or a light version of this style—which most people cannot distinguish in blind taste tests (Tremblay and Tremblay 2007). This uniformity is remarkable when you consider that that the Great American Beer Festival judges eighty-four different beer styles, many of which have a number of sub-categories (Hannaford 2007).

Not surprisingly, the remaining brewers became even bigger in the years after prohibition, as technological developments such as canning, refrigeration and faster transportation made geographic expansion easier. Initially, however, growth by acquiring other firms was mostly blocked due to strong antitrust enforcement. In 1959, for example, the tenth largest brewery in the United States, Pabst, acquired the eighteenth largest brewery, Blatz, resulting in a combined national market share of 4.5 percent. Regulators opposed the acquisition, but Pabst fought this decision all the way to the Supreme Court. In 1966, the majority ruled that the buyout had to be reversed, with Justice Hugo Black stating, "If not stopped, this decline in the number of separate competitors and this rise in the share of the market controlled by the larger beer manufacturers are bound to lead to greater and greater concentration of the beer industry into fewer and fewer hands" (Rosenbaum and Cox 2009). Less

Table 4.1 US beer market, 2012

Firm	Market share (percent)
Anheuser-Busch InBev	46.4
MillerCoors	27.6
Crown Imports	5.8
Heineken USA	4.0
	CR4: 83.8

Source: Beer Insights (2013)*
*These numbers underestimate market share, as nearly all of Pabst's beers, with 2.8 percent market share, are brewed under contract by MillerCoors. In addition, in 2012 AB InBev owned 50 percent of Crown Imports.

than five decades later, as Table 4.1 indicates, just two firms controlled approximately three-fourths of national sales.

Until the 1980s, when antitrust enforcement was relaxed, gaining increased market share came primarily through geographic expansion, as well as through the strategy of colonizing mind space. The largest firms had an advantage in this process, though, as they could afford to advertise nationally at lower per capita rates when compared to local ads (Tremblay and Tremblay 2007; George 2011). Anheuser-Busch's growth was almost entirely driven by advertising on these public airwaves; in contrast to competitors, their CEO in the 1980s and 1990s, August Busch III, made very few acquisitions (MacIntosh 2011). The firm is well known for its memorable ad campaigns, such as Spuds Mackenzie, Bud-weis-er frogs, Clydesdale horses, Bud Bowls and Whassup? guys, all of which significantly increased sales (Holt 2003), and helped increase market share from 28 percent in 1980, to 48 percent in 2007 (Brock 2011). Many of these ads were also appealing to children, but strong public criticism encouraged the firm to simply shift its campaigns from one anthropomorphic animal to another (Wallack, Cassady, and Grube 1990; Balu 1998).

More than a billion dollars a year is spent on advertising beer in the United States, primarily for television ads, and nearly 80 percent of this money comes from the top two firms (Mintel 2013c). Sporting events, from major international sports to local softball leagues, are also blanketed with beer billboards and banners. The exclusive marketing rights secured by the largest firms, for many sporting events, especially Anheuser-Busch, led to complaints from competitors and an antitrust investigation in the 1980s, but the practice was allowed to continue (MacIntosh 2011). A growing area of marketing for beer (and numerous other products) is paying for them to be featured in television and movies (Jernigan 2009). The film *Minority Report* set a record of $25 million in product placement fees in 2002, including payments from Diageo to highlight their Guinness brand beer (Grossberg 2002). This record has since been broken many times, and stands at $160 million for *Man of Steel*, including a reported $23.5 million to have Superman's alter ego, Clark Kent, drink Budweiser (Boshoff 2013).

Some beer firms focus entirely on marketing and leave the production to others. Pabst is one of these virtual brewers, which encompasses two dozen formerly independent beer brands. Nearly all of this beer is brewed under contract by SABMiller, yet investor C. Dean Metropoulos sold the firm for approximately three times his purchase price (from $250 to $750 million) in a space of just four years (Stanford 2014). The new owners in 2014 included the venture capitalist firm TSG Consumer Partners, which once controlled Vitaminwater before selling it to Coca-Cola. Boston Beer Company, which owns the Samuel Adams brand, started out brewing under contract but eventually developed its own breweries (So 2013).

Shelf space is another battleground for the beer industry, which is meticulously engineered through the use of planograms. These are visual representations of the placement of every brand and size of product in the category to fit the allotted shelf space, with a goal of maximizing sales. Product proliferation has enabled the largest firms to take up more space in these planograms. ABInBev's Bud Light, for example, comes in numerous, slightly different forms such as Bud Light Ice, Bud Light Platinum, Bud Light Lime, Bud Light Lime Straw-ber-Rita, and Bud Light Chelada.

Retailers frequently give their largest supplier in a given product category the responsibility to be the "captain," and design the placement of all these products, including those of direct competitors. These firms take great pains to point out that, in some cases, the sophisticated software they use to optimize sales also benefits competing firms (Kurtuluş 2014). Nonetheless, a Federal Trade Commission report found that: "The category captain might: (1) learn confidential information about rivals' plans; (2) hinder the expansion of rivals; (3) promote collusion among retailers; or (4) facilitate collusion among manufacturers" (Federal Trade Commission 2001; Lynn 2006, 50).

The dominant brewers' powers as category captains may be utilized against much smaller firms, known as craft brewers or microbrewers, which have grown in popularity since the 1980s. By 2014 there were more than 2,700 breweries in the United States, and the market was showing a pattern of divergence. On one side was an increasingly concentrated mass market with flat sales growth and on the other was a very fast-growing and increasingly competitive niche—craft brewers accounted for more than 7 percent of the market by volume and nearly double that by sales value in 2013 (Brewers Association 2014). These markets are also distinguished by the macro-brewers emphasis on pale US lagers versus a much wider (albeit more expensive) selection in the craft market, including ales, stouts, porters and sour beers. Even the most mainstream retailers are now likely to offer a selection of craft beers, particularly those produced by larger firms, such as Boston/Sam Adams, Sierra Nevada and New Belgium. The Boston Beer Company accounts for just 1.3 percent of sales by volume but enough to make founder Jim Koch a billionaire (Coffey 2013).

The big brewers' skills in product proliferation have enabled them to respond to this trend by jumping on the craft brewing bandwagon. Because they lack the legitimacy to link their own names to these efforts they use two different strategies to colonize this space. The first is to buy up existing craft brewers, as with Anheuser-Busch's acquisitions of Chicago's Goose Island in 2011, New York's Blue Point Brewing in 2014, and a one-third stake in Craft Brewers Alliance or MillerCoors acquisitions of Leinenkugel's and Henry Weinhard's. The second strategy is to introduce fake craft beers, as with Anheuser-Busch's brands Shock Top, Wild Blue, and Landshark, or MillerCoors' brands Blue Moon and Third Shift (see Figure 4.1).

Global expansion in the industry also involves a form of stealth ownership, as buyouts of giant US brewers by their foreign competitors are kept relatively hidden. Anheuser-Busch was acquired by InBev (based in Belgium but managed at the top levels by executives from Brazil) in a hostile takeover in 2008 and renamed Anheuser-Busch InBev. The US firm was vulnerable to this action, despite being a very profitable business and accounting for nearly half of national sales, because it had not grown as rapidly as foreign competitors in other global markets (MacIntosh 2011). SABMiller, originally a South African company (Box 4.1) but now based in London, increased its US market share with two key moves: it acquired the second largest US firm Miller in 2002, followed by a joint venture with the smaller US/Canadian firm, MolsonCoors, in 2007. Heineken and Carlsberg are currently the third and fourth largest firms globally but not as dominant outside of Europe and Asia. As these four European-based firms have consolidated control of approximately half of the global beer industry in recent decades, they learned that the majority of beer drinkers have strong cultural attachments to local or national brands. Consequently, the parent companies have not drawn much attention to their alliances and acquisitions. AB InBev, for example, continues to run Budweiser

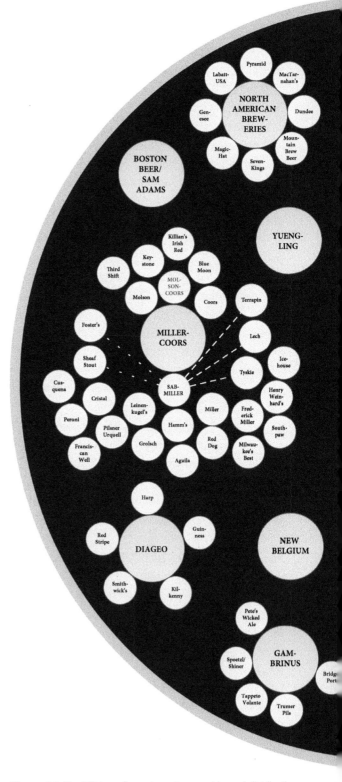

Figure 4.1 Top US beer firms: brand ownership and distribution.

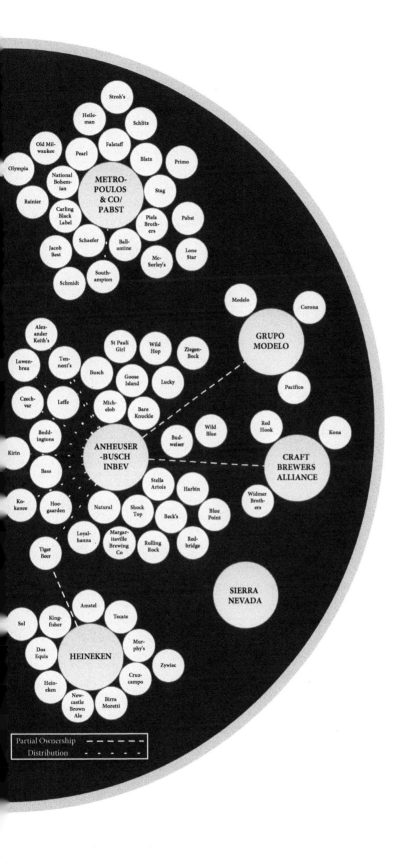

METRO-POULOS & CO/ PABST

Stroh's
Heileman
Schlitz
Old Milwaukee
Pearl
Falstaff
Blatz
Primo
Olympia
National Bohemian
Stag
Rainier
Carling Black Label
Piels Brothers
Pabst
Jacob Best
Schaefer
Ballantine
Lone Star
Schmidt
Southampton
McSorley's

GRUPO MODELO

Modelo
Corona
Pacifico

ANHEUSER-BUSCH INBEV

Alexander Keith's
St Pauli Girl
Wild Hop
Ziegen-Bock
Lowenbrau
Tennent's
Busch
Goose Island
Lucky
Czechvar
Leffe
Michelob
Bare Knuckle
Boddingtons
Budweiser
Wild Blue
Kirin
Bass
Kokanee
Hoegaarden
Natural
Stella Artois
Harbin
Shock Top
Beck's
Blue Point
Tiger Beer
Loyalhanna
Margaritaville Brewing Co
Rolling Rock
Redbridge

CRAFT BREWERS ALLIANCE

Red Hook
Kona
Widmer Brothers

SIERRA NEVADA

HEINEKEN

Amstel
Tecate
Sol
Kingfisher
Murphy's
Dos Equis
Zywiec
Heineken
Cruzcampo
Newcastle Brown Ale
Birra Moretti

Partial Ownership — — —
Distribution - - - - - -

Box 4.1 Dodging Taxes: SABMiller

One way to increase profits is to pay less taxes or avoid paying taxes altogether, and large firms are more easily able to exploit loopholes that make this possible. In 2010, the UK nonprofit ActionAid accused SABMiller of shirking its tax burden around the world. They noted that the corporation had more tax haven companies (65) in Africa than breweries and bottling plants, and that this strategy was estimated to deprive India and African countries of $31 million a year. Because they paid no taxes at all in Ghana, the report noted in reference to a woman who sells beer and food from a stall near one of SABMillers's breweries: "astonishingly she has paid more income tax in the past two years than her neighbour and supplier, which is part of a multi-billion pound global business" (Hearson and Brooks 2012, 7).

SABMiller is not unique, as eighty-three of the top 100 firms headquartered in the United States have subsidiaries in countries that are considered tax havens, such as the Cayman Islands, Bermuda, and Luxembourg (McIntyre et al. 2011). They use a strategy called transfer pricing to move transactions through paper companies and offshore banks and shift profits away from high-tax countries. More specifically, SABMiller used strategies such as: (1) placing the ownership of fully African brands (such as Castle, Stone, and Chibuku) in the Netherlands, so that the millions paid in royalties by their African subsidiaries were taxed at a very low rate by the Dutch government, (2) directing African and Indian subsidiaries to pay large "management service fees" to sister companies in tax havens such as Switzerland, and (3) routing goods through the very distant nation of Mauritius, where they could be taxed at 3 percent, rather than the 25 percent rate levied in Ghana (Hearson and Brooks 2012, 8). Although all of these strategies were technically legal, following this report a number of countries in Africa met to discuss tax avoidance issues. The Organization for Economic Co-operation and Development (OECD) soon followed with a project called "tax inspectors without borders," with a goal of challenging transfer pricing, particularly as it affects less industrialized countries (Crotty 2013).

commercials with patriotic themes, package their products to resemble the US flag, and list the manufacturer as "Anheuser-Busch, Inc., St. Louis, Missouri, U.S.A." (Howard 2014b). They also sponsor an annual "Made in America" music festival, which started in 2012.

This increased concentration has not been beneficial for US consumers. Soon after InBev acquired AB in 2008 they moved to raise prices, and in 2011 increased them again (Frankel 2010). In addition, they have reduced the quality of the product by seeking out cheaper ingredients, and allegedly watering down alcohol content (Tuttle 2013). Boosting profits at the expense of quality is a mistake that the Milwaukee, Wisconsin brewer Schlitz made in the 1970s, as consumers noticed the changes and their sales plunged (Van Munching 1997). Because there were more mass-market alternatives at the time, the firm never recovered and was subsequently bought out by Stroh's (which itself was later acquired by Pabst). AB InBev has similarly experienced declining sales in the United States but has reassured investors by continuing to increase profits in other areas (Peterson 2012).

Interestingly, the DOJ took action when AB InBev moved to take over half of Mexico's beer market (and its US imports), via a $20.1 billion buyout of Grupo Modelo. The DOJ made note of the recent price increases, which were quickly matched by MillerCoors in a classic example of tacit coordination (Kendall and Bauerlein 2013). Because these two firms already control the vast majority of the US market, average beer prices have increased much faster than wine and spirits in recent years (Gray 2013). One firm that resisted the price increases, however, was Grupo Modelo. As a result, AB InBev was allowed to acquire the Mexican firm but forced to make a concession of licensing US sales to another alcohol giant, Constellation Brands, as well as to sell them Grupo Modelo's huge, state-of-the-art brewery, located just across the Texas border (Scott 2013). Price fixing involving the largest beer firms has been more visible in Europe, where there are more competing firms to blow the whistle to regulatory agencies. For one case that took place from 1996 to 1999, the fines amounted to $370 million (Associated Press 2007).

A new market: soy milk

Steve Demos had toiled since the 1970s to increase Americans' consumption of soy products. The firm he founded, WhiteWave, had little success with his first product, tofu, nor with subsequent products, which included tempeh, miso, and soy ice cream (Dobrow 2014). He experienced skyrocketing sales in the late 1990s, however, with his innovation of moving soy milk from shelf stable packaging to cartons in the refrigerated dairy case, under the brand name *Silk*. Silk's share of the soy milk market rose from zero in 1996 to approximately 85 percent by 2003, for example, when it could be found in 98 percent of grocery stores (Silverstein 2011). Soy milk sales had been increasing slowly since the

1980s when Hong Kong-based Vitasoy entered the US market (SANA 2014), but Silk's new shelf space location made the product much more visible to mainstream consumers. The success of this spatial colonization strategy led to similar marketing efforts for other milk substitutes, eventually cutting into the sales of soy milk.

Table 4.2 shows the leading brands and market share for milk substitutes. Soy milk claimed 7 percent of combined milk and milk alternative sales in 2012, but almond milk was gaining with a 6 percent share and coconut milk had a 5 percent share (Mintel 2013a). These products are facing increasing competition from other milk alternatives, including rice, oat, hazelnut, sunflower, flaxseed, and hempseed milks. The soy milk market by itself was worth $846 million in 2011, declining from a peak of $981 million in 2009 (Haderspeck 2012). This market remains quite concentrated though, with WhiteWave claiming a 74 percent share of soy milk sales (Salerno 2014). The firm responded quickly to growing interest in other milk alternatives and currently has dominant market shares in coconut and almond milk sales as well.

Gaining access to the dairy case, particularly in conventional supermarkets, required paying slotting fees, or charges that were levied by most retailers in exchange for being placed on store shelves (Aalberts and Jennings 1999). As noted in Chapter 2, retailers are quite secretive about these fees, but they are estimated to start at tens of thousands of dollars just to access a regional chain for one product (Glaser and Thompson 2001). Demos did not have the resources to buy his way into mainstream outlets, so he approached Coca-Cola, Kraft and other enormous firms for assistance. He succeeded in convincing Dean Foods to acquire a minority stake in WhiteWave and to help with manufacturing in order

Table 4.2 US milk substitute market, 2012

Firm (Product)	Market share (percent)
WhiteWave (Silk soy milk)	36.6
Blue Diamond (Almond Breeze)	16.8
Lifeway (kefir)	5.3
8th Continent (soy milk)	3.2
WhiteWave (Silk coconut milk)	2.6
So Delicious (soy milk)	1.6
	CR4: >64.5

Source: Lazich and Burton (2014).

to expand distribution from the natural foods channel to the much larger super-market channel (Adamy 2005).

In the late 1990s, Dean Foods CEO Gregg Engles was overseeing the consolidation of the US dairy industry, and beginning to test the waters of the rapidly growing markets for organic milk and milk alternatives. Engles' previous careers included practicing law, followed by consolidating the packaged ice business—his first move was to acquire Reddy Ice from 7-Eleven (Martin 2012). After buying Suiza and dozens of other conventional dairy firms, including Dean Foods (which became the name of the new dairy giant), he acquired minority stakes in Horizon, a large organic dairy, as well as in WhiteWave. By the early 2000s Dean had purchased 100 percent equity in both. In addition to the strategy of shelf space, resources were increased for mind space, such as spending $22 million on an ad campaign for Silk (Adamy 2005). These investments paid off with rapid sales growth and extremely dominant market shares.

There was a downside to this rapid growth, however, which at one point was measured in *weekly* sales increases of 3–5 percent (SANA 2014). When Dean Foods exercised the option to increase its stake in WhiteWave to 100 percent, it was initially blocked by Demos, who did not think Engles shared his idealism (Adamy 2005). Dean won the lawsuit but allowed Demos to stay on as an executive for a brief time before being forcing him out entirely (Fromartz 2006). Dean then made several cost-cutting, profit-boosting moves (Box 4.2), first moving away from domestically sourced organic

---

Box 4.2 Cheapening Ingredients: 8th Continent/Stremick's

8th Continent is a soy milk brand that states prominently on the carton that it is made from US-grown soybeans. How are they able to do this and still retail for less than their larger competitor, Silk? One reason is that they have found other ways to cut production costs. A food scientist at Nestlé explained the unrelenting pressure that dominant food manufacturers face in their efforts to beat the average rate of profit: "we were always trying to make it cheaper" (Moss 2013, 155). In 8th Continent's case, part of the solution is to use more ultra-processed ingredients, genetically engineered soybeans, and artificial flavors and sweeteners. They utilize isolated soy protein and soybean oil rather than whole soybeans, for example. Although 8th Continent once stated on the packaging that they source non-GMO soybeans, that claim was dropped in 2013. Silk has moved in the other direction, obtaining third-party certification from the Non-GMO Project in 2010. 8th Continent also differs from competitors by using artificial flavors manufactured in chemical plants and the artificial sweetener sucralose.

Because most of these ingredients are on the banned lists at leading natural foods retailers, 8th Continent is typically found only at conventional retailers, such as Walmart. Its low price strategy, including frequent promotions and discounts, and a taste engineered to be preferred over competing soy milk products, have helped to increase sales exponentially (Garrison-Sprenger 2003). The brand originated as a joint venture between General Mills and DuPont, in 2000, as a way to market Solae soy ingredients—this firm was originally a joint venture between DuPont and grain trader Bunge but is now fully owned by DuPont. 8th Continent was sold to a conventional dairy firm, Stremicks's Heritage Foods, in 2008 but maintains a long-term supply agreement with Solae (DuPont 2008).

soybeans and buying them more cheaply from China and Brazil, followed by stealthily converting many of their products from organic to "natural" in 2009. The familiar packaging, UPC symbols and prices remained the same, but the product inside was reformulated with much cheaper conventional, imported soybeans. The only clue to the change was that the word "organic" was dropped from the product description and ingredient list. A number of retailers failed to notice this, as they had not been informed, and incorrectly continued to advertise these products as organic (Shlacter 2009). Before the acquisition, Silk was 100 percent organic, but according to the watchdog group, Cornucopia Institute, by 2014 just 6 percent of their products carried this designation (Cornucopia Institute 2014). Demos has been critical of the actions occurring since he left the soy industry, stating: "There have been some significant changes to the 'authenticity' of what was once the (nation's) largest organic brand" (Wallace 2013).

The conventional side of Dean Foods did not grow as quickly as the natural/organic foods subsidiaries, and by 2010 Engles was facing strong criticism from Wall Street analysts—his compensation averaged more than $20 million a year from 2006 to 2011 even though the stock price declined an average of 11 percent annually during this time (Williams 2012). As a result, in 2013 Engles stepped down from Dean to become CEO of a separate, publicly traded firm, created by spinning off all of the natural/organic brands that had been acquired. Perhaps to Demos' chagrin, the new firm was named WhiteWave. Engles immediately announced his intention to make more acquisitions to consolidate the natural/organic space even further. One of his first steps was to acquire a pioneer in organic bagged salads, Earthbound Farm, for $600 million. This California-based processor controlled the most dominant packaged organic salad brand, accounting for 55 percent of the category (O'Halloran 2014). With this move, WhiteWave became the world's largest natural/organic food firm, achieving a market capitalization in 2013 of approximately $5 billion, much larger than Dean Foods' conventional dairy business at $1.5 billion. By contrast, Hain Celestial, which grew through acquisitions of several dozen natural/organic food and cosmetic brands since the 1990s, had a market capitalization of just $4.2 billion.

A repackaged market: Bagged salad

Earthbound Farm played a significant role in continuing the deskilling of US consumers by encouraging consumption of pre-washed, pre-cut, bagged salads. In the process, the firm helped expand a market that barely existed before 1990 to sales of approximately $4.1 billion annually by 2012 (Mintel 2013b). It also contributed to revitalizing produce giants Dole and Chiquita (Box 4.3), which were facing declining sales of the essentially unbranded commodity of lettuce. Sales of bagged salads are not only increasing, they are generating greater profits by incorporating a small amount of additional processing.

Precut salads were developed for restaurants and institutional food services in the 1970s, but Drew and Myra Goodman, the founders of Earthbound Farm, were among the first to begin selling packages in household sizes. The Goodmans started farming organic raspberries in 1984, and then expanded into leafy greens. When a key restaurant customer went out of business in 1986 they faced a dilemma of what do with their excess baby lettuce. Because they washed and bagged lettuce for their own personal consumption, they had the idea to sell it to a San Francisco natural foods store in much smaller packages. Organic salad mix made from an assortment of young leafy greens was already popular with upscale restaurants, following trends set by Alice Waters of the French-inspired Chez Panisse in Berkeley, California. Waters described her mix as *mesclun*, which contributed to a gourmet image and premium prices and led the organic farmers enlisted in meeting the rising demand for this product to call it "yuppie chow" instead (Guthman 2003). Earthbound Farm continued to leverage its advantage as the first entrant in this market to grow exponentially, eventually controlling 70 percent of US sales of organic bagged salads, primarily through conventional retailers (Carson Private Capital 2009).

The success of organic salad mix quickly attracted imitators for conventionally grown leafy greens, and Dole and Fresh Express were able to pay the slotting fees required to dominate much larger mainstream markets. Fresh Express was first acquired by the distributor Performance Food Group but then sold to Dole's biggest competitor in fruit and vegetable production, Chiquita Brands. In 2005, when the deal was completed, Fresh Express controlled approximately 40 percent of the bagged salad market. The $855 million cost of the acquisition was repaid quickly, with higher-margin, value-added bagged salads boosting Chiquita to record sales (Chiquita 2006).

Firms that do not pay slotting fees are likely to lose the account, yet even after paying them, there is little guarantee of commitment on shelf space. One way to avoid fees, however, is to become a supplier of the retailer's private label bagged salads (Glaser and Thompson 2001). Dole, for example, packages private label bagged salad for retailers, including Kroger and Walmart (Ordman 2012), and Earthbound Farm does the same for Safeway, Trader Joe's and Loblaws (DeLind and Howard 2008). Table 4.3 shows the current market share for leading firms. They have lost branded market share compared to previous years, when the top two were estimated to control more than three-quarters of all sales. Yet because of the relatively low margins in the industry and reduction in slotting fees, they may be more profitable as a result (Mintel 2013b).

Bagged salads may not seem like an impressive example of deskilling, because it is relatively easy to (re)acquire the knowledge of how to wash

Table 4.3 US bagged salad market, 2012

Firm	Market share (percent)
Chiquita/Fresh Express	32.2
Dole	22.2
Earthbound Farm	5.8
Ready Pac	4.4
	CR4: 64.6

Source: Cook (2012) from Neilsen*
*This figure underestimates true market share because the percentages refer only to branded products, and Dole and Earthbound also packaged private label bagged salads (Fresh Express announced it would also enter the private label market in 2012).

and chop greens. On the other hand, perhaps there is no better example of how firms have convinced growing number of people to avoid even the most minimal amount of work in exchange for higher prices. Retailer shelf space for such products continues to increase, such as cake mixes that replace the need to measure a few dry ingredients, pre-scrambled eggs, and peanut butter and jelly in the same jar. Perhaps the most dramatic example was offered by the Billa retail chain in Austria, before public criticism put a stop to it: pre-peeled bananas, placed on a foam tray and wrapped in cellophane (Pous 2012).

By relying on dominant firms to do most of the work of preparing salads, consumers likely pay a price in reduced nutritional value (Robinson 2013). The more immediate risk, though, is food safety. Bagged salads have been implicated in a number of foodborne illness epidemics, which have resulted in deaths and permanent disabilities (e.g., kidney failure). In 2006, for example, suspected contamination from a single spinach farm led to an epidemic of *E. coli* O157:H7 that affected twenty-six states and led to a recall of more than forty brands (DeLind and Howard 2008). The highly concentrated production of these foods means that even if problems are rare, when something does go wrong it could affect a much greater number of people than with a more decentralized system. Product recalls have also resulted from the detection of pathogenic salmonella, listeria, and cyclospora. Disease outbreaks linked to leafy greens have increased to account for nearly a quarter of all foodborne illnesses, and those most at risk from exposure to these pathogens include the elderly, the very young, and pregnant women (Gallant 2013). Food scientists suggest that buying unprocessed greens and preparing them at home is much safer, and even some growers for these firms admit they will not eat bagged salads (Stuart 2011).

The industry continues to offer more convenient and expensive forms of leafy greens to increase profits. These include single serving size salads or larger "kits" with additional ingredients, such as dressings, meats, cheese, fruits, and grains. Bagged salads are particularly popular with younger consumers, and 63 percent of those under age twenty-four report buying them at least monthly (Kroger 2011). Some firms are using social media platforms, such as Facebook, Twitter, and Instagram, to inexpensively increase mind space, and capture more market share. Ready Pac, for example, which dominates the single-serve salad segment with 85 percent of sales, held an Instagram photo contest ($500 prize) to promote their plastic clamshell salads (Hornick 2013).

Box 4.3 Overthrowing Governments: Chiquita and Dole

The term "banana republic" is often associated with what are now the top two firms in the bagged salad industry, Chiquita and Dole. These diverse fruit and vegetable firms were previously known by different names: United Fruit became Chiquita, and Standard Fruit became Dole. United Fruit's monopolization of the banana industry in Honduras, and its close ties to the autocratic government of this country in the early 1900s, inspired the author O. Henry to describe a fictional nation as a "banana republic" (Henry 1904, 132). The firm's domination of politics in a number of Latin American countries continued, however. This included convincing the CIA to overthrow the elected government of Guatemala in 1954 after the president, Jacabo Arbenz, implemented land reforms—the US government had already invaded Latin American nations more than twenty times in previous decades to protect business interests (Bucheli 2008). Chiquita was also suspected of playing a role in a coup in Honduras in 2009 (Kozloff 2009).

Both United Fruit and Standard Fruit interrupted shipments from several Latin American countries and threatened to fire workers in the mid-1970s, when reform movements increased taxes and eliminated generous subsidies (Bucheli 2008). More recently, both successor firms were accused of funding guerrilla/paramilitary groups in Colombia, such as FARC (Revolutionary Armed Forces of Colombia), AUC (United Self-Defense Forces of Colombia), and ELN (National Liberation Army). These groups were responsible for killing thousands of people, including small farmers and union leaders who challenged the firms' interests. Chiquita admitted making payments of $1.7 million over a seven-year period but argued that this was due to extortion. The corporation pleaded guilty to violating anti-terrorism laws in 2007 and was assessed a $25 million fine (Maurer 2009). Dole, in contrast, did not admit to making payments but sworn statements from former paramilitary group members claim that funds were provided by both firms without coercion (Cray 2010).

Conscious consumers?

While dominant firms' marketing has traditionally focused on a one-way flow of information via mass media outlets, the innovation of social media has made it easier for customers to communicate in the other direction, as well as with each other. This has enabled activists to put more pressure upon the world's largest food and beverage corporations, and in some cases, even change their practices. Pepsi and Coca-Cola, for example, pledged to remove the controversial ingredient brominated vegetable oil from their beverages after an online petition started by a teenager from Mississippi received hundreds of thousands of signatures (Choi 2014). In addition, Kraft pledged to remove artificial yellow dyes first from some of their varieties of macaroni and cheese and later from all of them, due to a similar effort (Wilson 2013; Gillam 2015). The stereotype of consumers as unconscious dupes is challenged by these cases, showing that a vocal minority can trigger some positive changes, even if they do not significantly threaten corporate power (Goodman, DuPuis, and Goodman 2012; Johnston and Cairns 2013).

Public advocacy groups have also been effective in reducing the use of ultra-processed ingredients. Hydrogenated oils (often soy or cottonseed oils) have high levels of trans fats, which have negative impacts on cardiovascular health. The federal government responded to pressure to limit consumption of these ingredients by mandating labeling in 2006, and some state and local governments have banned their use in restaurants and schools. The increased negative public perception of hydrogenated oils reduced sales of foods containing these ingredients—including crackers, cookies, chips, and cereal—and encouraged food manufacturers to find substitutes (Stobbe 2012). Some public health scientists suggest these changes may have contributed to a surprising recent decline in children's cholesterol (de Ferranti 2012). High fructose corn syrup has also been targeted by health activists, and manufacturers, such as Kraft and Sara Lee, are beginning to remove this ingredient from some of their processed foods as well (York 2010).

Children continue to consume highly processed foods in large amounts, and efforts to restrict marketing to children have been less successful. Many leading food and beverage firms have agreed to voluntary restrictions, but these have had little to no impact on children's exposure (Galbraith-Emami and Lobstein 2013). While media literacy programs attempt to reduce children's susceptibility to such exposure, they are unlikely to apply this knowledge outside of the classroom (Rozendaal et al. 2011). In addition, efforts to inspire greater awareness of marketing techniques in adult populations, such as the anti-ads in *Adbusters* magazine, are often co-opted by marketers, serving to reinforce their sales efforts (Rumbo 2002).

Another strategy of resistance is to avoid purchasing processed foods and engage in reskilling efforts. This do-it-yourself (DIY) movement is particularly strong in beer brewing. More than one million people in the United States brew beer at home, the majority of whom are young, college educated men (Murray and O'Neill 2012). In addition, regulatory barriers continue to be reduced—Mississippi and Alabama were the last remaining states to relegalize home brewing in 2013 (Chappell 2013). Making soy milk or other milk alternatives at home is also popular, in part due to the cost savings, either with specialized appliances or simply blenders and strainers (Brewer 2011).

More than two-thirds of states have passed cottage food laws in recent years, which allow people to sell processed foods that they make at home. States without such laws require food sold to the public to be made in commercial kitchens, which is a significant barrier to many potential entrepreneurs. Because sales for cottage food operations are frequently capped well below the average household income, they present little threat to giant food processors but may provide an opportunity for people to bootstrap their way to larger businesses (Condra 2013). These laws are also one of the few recent examples of regulations changing in a direction that encourages more competitive, rather than less competitive markets.

Although from a distance it appears that there has been little success in slowing the increasing power of packaged food firms, a closer look reveals the enormous engineering required to continuously steer our consumption patterns in directions preferred by these firms. Leading beer and soy milk firms are experiencing declining sales in key markets and have emphasized expansion into new products and geographic regions in their attempts to beat average returns. Leafy greens firms have successfully repackaged their products as pricier bagged salads but also opened themselves to greater liability, such as paying out tens of millions of dollars to settle lawsuits brought by people who fell ill from consuming their products (Tesoriero and Lattman 2006).

Consumers are also not their only challengers. Most of these firms rely on upstream suppliers for key ingredients, a dependency that the latter can potentially exploit by manipulating prices. The following chapter describes how the less publicly visible firms involved in the collection and primary processing of farm products are able to influence prices, frequently at the expense of both customers and suppliers.

Chapter 5

Manipulating prices: commodity processing

They'll negotiate; they're corporate.
—Johnny Mnemonic (*Johnny Mnemonic*)

Commodities are products that are undifferentiated and can easily be substituted from a number of different sources. In theory, their prices should be set by the market as a whole, as there are no unique attributes to discourage buyers from seeking out the lowest price. As a small number of firms have increased their control over key commodities, however, they have gained the power to manipulate prices to their advantage, such as driving down the amounts they pay to suppliers and driving up the amounts they charge to buyers. This power is exercised using both legal and illegal means.

This chapter analyzes concentration in commodity processing industries and its impacts, with a focus on price manipulation in soybeans, dairy, and pork. These strategies have been aided by decreasing transparency in these markets, with fewer opportunities for the public to uncover who is benefiting and who is losing from exchanges of these commodities. The rising power of dominant firms has contributed to lower shares of retail prices for many farmers, for example, even though farm expenses have increased (Moeller 2003; National Farmers Union 2014b). Other consequences of consolidation include steep drops in not only the ownership of processing facilities but in the number of processing facilities and the number of people employed at these operations. In addition, processing has become more geographically concentrated, increasing both its environmental impacts and the distance required to reach end markets (Jackson-Smith and Buttel 1998; Welsh, Hubbell, and Carpentier 2003).

The customer is our enemy

One of the most famous examples of an illegal price-fixing scheme was recorded by the US Federal Bureau of Investigation (FBI) in the 1990s. It involved the commodity processor Archer Daniels Midland (ADM), and its Japanese and Korean "competitors" in the markets for citric acid and lysine—these are additives in food or animal feed. Executives at ADM famously had a saying that "the competitor is our friend and the customer is our enemy" (Lieber 2000, 10). Key evidence was obtained when a high-level ADM executive, Mark Whitacre, agreed to wear a wire for the FBI. He recorded conversations in hotels in Hawaii and Mexico City, in which the firms established sales volumes and agreed on higher prices. Other food and beverage firms paid hundreds of millions of dollars more than they would have without these price-fixing agreements, which ultimately resulted in greater costs for consumers (Connor 2007).

Major US media outlets largely ignored the scandal, particularly the Sunday morning political talk shows, nearly all of which were sponsored by ADM (Farah and Elga 2001). The power of the firm was also demonstrated in the jail sentences handed out to three of their executives: the whistleblower, Whitacre, received the longest term. ADM did receive the largest antitrust fine ever levied to that point, $100 million, but this was less than the maximum allowed, and likely less than was earned in monopoly profits during the conspiracy (Connor 2007). The firm was also investigated for price fixing in the much larger high fructose corn syrup market, but granted immunity in exchange for accepting the fines for lysine and citric acid. This case became the subject of two books—*Rats in the Grain* by James Lieber, and *The Informant* by Kurt Eichenwald—as well as a Hollywood movie, starring Matt Damon. Eichenwald's book was more popular than Lieber's, but it was far less critical of ADM executives and the government prosecutors.

Many other firms in this stage of the food chain are able to manipulate prices through more subtle means, such as price signaling or owning enough of upstream stages of production to exert control over market prices. Although the US antitrust laws of the early twentieth century were designed to prevent these behaviors, changes in interpretation and enforcement have enabled them to once again become effective tools for dominant commodity processors. Because this stage is more hidden from view than other parts of the food system, many consumers are unfamiliar with the names of these firms, as well as their increasing dominance. Even the firms that are more vertically integrated tend to offer so many branded products that the extent of concentration is obscured.

The ABCs of profit: soybeans

The four firms that dominate worldwide trading and processing of soybeans, as well most important grains, are known as ABCD companies, an acronym formed by the first letters of their names: ADM, Bunge, Cargill and (Louis) Dreyfus. Not all grain is traded over borders—only one-third of soybeans leave their country of origin, for example—but these firms are estimated to control as much as 90 percent of the global grain trade (Murphy, Burch, and Clapp 2012). Unlike dominant firms in other parts of the food system, such as retailing or seeds, these grain processors have a very long history; all were founded in the 1800s or early 1900s (Clapp 2014). They have many other similarities as well, including control by private families and a strong tradition of secrecy as a means to increase their power.

The power of most of these firms in the late 1800s resulted in a number of responses in the US, including greater government control of trading monopolies, establishing price floors for commodities, and the formation of farmer cooperatives as alternatives (Murphy 2006). Cooperatives sought to increase control over market prices by forming associations, pooling their grain, building elevators for storage and transporting their products to terminal markets. At their peak, in the 1920s, there were more than 4,000 elevator cooperatives nationally (Schneiberg, King, and Smith 2008). The majority of these cooperatives eventually went out of business, and some of the remaining organizations have frequently engaged in mergers to increase their scale and keep pace with industry changes. Ag Processing Inc., for example, was the fourth largest soybean processor in the United States in 2011. In addition, some of these cooperatives have joint ventures with corporations, such as ADM and Cargill (Gray, Heffernan, and Hendrickson 2001), as do some of the largest farmer cooperatives in the EU (Heffernan, Hendrickson, and Gronski 1999).

By the 1980s, the top five grain trading firms continued to be controlled by just seven families (Morgan 1979). They increased the pace of buyouts of competitors, and from 1982 to 1992 an estimated 453 oilseed processing plants were acquired by the largest firms (Brock 2011). From 1972 to 1992 the CR4 in the United States increased from 52 percent to 80 percent, and the number of employees fell by 43 percent (Ollinger et al. 2005). One of the most significant acquisitions was Cargill's takeover of Continental in 1998, which resulted in the combination of the first and second largest grain exporters in the United States at the time. The Department of Justice initially opposed the deal, but in a typical pattern, it was eventually allowed to go through with some concessions—these included divesting some ports in regions where they would have obtained a near monopoly (O'Brien 2004).

Soybean processors had a crush capacity CR4 of approximately 80 percent in 2011 (Table 5.1), and the estimated CR4 for market share was slightly higher, at 85 percent (James, Hendrickson, and Howard 2013). The leading firms are also dominant in the processing, transport and trading of corn, wheat, sugar, palm oil and cocoa, as well as products that result from further processing of these crops, such as food additives. ADM and Cargill also control much of soybean exporting from the US, and together with the firm Zen Noh, their market share is approximately 65 percent (Heffernan and Hendrickson 2002). Louis Dreyfus, based in France, is one of the largest firms globally but has smaller operations in the US.

Grain processors have a history of being able to use their size and information gathering resources to give them market advantages (Morgan 1979). This resulted in limits on trading in some exchanges, such as the Chicago Board of Trade (CBOT), in order to prevent excessive speculation (Clapp 2014). These limits are being relaxed, however, as powerful processing firms are now engaged more heavily in financial transactions to increase their profits. For many of these firms, their hedge funds and other financial instruments are a larger source of profits than their physical activities, a change that has been described as *financialization* (Clapp 2014; Isakson 2014). They are able to make money whether commodity prices move up or down, for example, if they are positioned to take advantage of regional differences or accurately anticipate the direction of movement. Even previous limits in commodity trading were less stringent compared to other financial markets, however, essentially allowing insider trading to be legal (Khan 2013b). This has attracted interest from more wealthy speculators and increased pressure

Table 5.1 Leading US soybean processors by crush capacity, 2011

Firm	Share of US crush capacity (percent)
Bunge	25.5
ADM	21.4
Cargill	21.2
Ag Processing	11.7
	CR4: 79.8

Source: US Soybean Export Council (2011).

to make these transactions even more opaque, which presents numerous problems for regulators (Vander Stichele 2012).

Traditional financial firms, such as Goldman Sachs, and traders of other commodities have been moving into agricultural markets in recent years. One example is Glencore, which began in mining and metals trading, but has increased its positions in grains, and is now the world's largest commodity trader. Their top employees may be paid much more than their counterparts at Wall Street firms, including annual bonuses of up to tens of millions of dollars (Schneyer 2011). These new competitors are using technologies to erode the "information oligopoly" that the ABCD firms have long enjoyed; additional threats to the traditional firms' dominance result from increasing restrictions on foreign investment in Asian countries, including China and South Korea, due to concerns about their dependence on these companies (Murphy, Burch, and Clapp 2012). The ABCD firms have responded by acquiring minority stakes and/or forming joint ventures with some of these new competitors. ADM, for example, owns 16 percent of the Asian firm Wilmar, and the two have joint ventures related to oil processing, fertilizers, and shipping.

One of the simplest means to use large size to manipulate prices to a firm's advantage is to corner the market or create a "squeeze." In the 1930s, for example, Cargill bought up corn futures after a crop failure in order to gain control of a large portion of the supply, which allowed the firm to increase prices. Cargill was subsequently expelled from the CBOT, and operated through independent traders for the next twenty-five years (Morgan 1979). In 1989, an Italian agribusiness firm, Ferruzzi, was suspected of trying to corner the market in soybeans when it accumulated futures contracts for 23 million bushels—more than twice the amount available for others, such as Cargill and ADM, to make good on these contracts. The CBOT responded by imposing a 1 million bushel limit on obligations to buy or sell soybeans (Atlas 1989). Ferruzzi sued for millions in lost profits and claimed that their actions were technically legal at the time, but ultimately: (1) dropped the suit, (2) transferred its CBOT membership to a US subsidiary, and (3) agreed to pay fines and legal fees of more than 3 million dollars (Bonanno and Constance 2010). The firm subsequently went bankrupt, and its US subsidiary was taken over by Cereol, which was itself later acquired by Bunge.

Another increasingly common means of controlling the soybean trade is through contracts with growers, which allows these firms to exert more influence over prices, and even farmers' decisions about what to grow (Khan 2013b). This strategy results in bypassing spot markets, for example, and reduces opportunities for price discovery (Murphy, Burch, and Clapp 2012). Where previously there were numerous buyers and sellers and frequent reporting of prices at numerous stages of the food system, now there are networks of firms that control the majority of these transactions via direct ownership, alliances or contracts (Hendrickson and Heffernan 2002). This raises concerns that more markets will be replaced by a "seamless system" that extends from the seed to the supermarket shelf, with costs kept completely secret (Heffernan 1999).

The global dominance of the ABCD firms also allows them to benefit from regional differences in government subsidies or tax breaks, as well as weaker labor or environmental regulations (Vorley 2003). Subsidies in the US, for example, enable soybeans to be purchased below their cost of production (Starmer, Witteman, and Wise 2006). The majority of commodity trades originate from the tax haven of Switzerland, and firms also commonly use transfer pricing to further reduce taxes. All four ABCD companies were accused of tax evasion by the government of Argentina, after reportedly submitting false declarations of sales in neighboring countries beginning in 2008, when soybean prices and processor profits increased dramatically (Murphy, Burch, and Clapp 2012).

Despite some conflicts with governments over tax avoidance, the ABCD firms have tremendous political influence in many countries. Louis Dreyfus is considered a cooperative in France, due to partial ownership of a farmer cooperative, which allows the firm to access credit at much lower interest rates than private firms (Vorley 2003). In the United States, former ADM CEO Dwayne Andreas viewed donations to candidates from both political parties as "tithing," (Box 5.1) and one donation involved him personally delivering $100,000 in unreported cash to Richard Nixon's secretary in 1972—the money sat in a

---

Box 5.1 Grainwashing: ADM, Bunge, and Cargill

"Greenwashing" is a term used to describe corporations that take numerous actions to harm the environment, but also devote a tiny percentage of their resources to addressing environmental problems. "Grainwashing" has similar connotations but refers to agribusinesses that structure society in ways that tend to increase hunger, while devoting a small fraction of their resources to the problem of food insecurity (Scanlan 2013). ADM, Bunge, and Cargill are some of the numerous commodity traders and agricultural input producers that use the rhetoric of "feeding the world" to distract the public from the political economy of food and agriculture. This rhetoric is frequently paired with population projections that predict 9 billion people by 2050 and suggestions that food production would need to be increased by 70 percent or more to adequately feed everyone. The latter figure is based on questionable assumptions, such as continued high rates of food waste and spoilage and inadequate attention to regional and local variations (Wise 2013).

Most famines are not caused by insufficient production of food but social factors related to its distribution, particularly the inability of poor people to pay for the means of subsistence (Sen 1990). Although the prevalence of hunger has declined slightly in recent decades, nearly one out of eight people globally experience chronic undernourishment (de Schutter 2014). This results in hunger-related deaths of more than 8,000 children every day (Black et al. 2013). In the United States, which produces far more food than its population can consume, one in seven people report relying on charitable donations to feed their households (Feeding America 2014).

Even programs ostensibly designed to address the problem of hunger tend to benefit ADM, Bunge, and Cargill far more than the poor (Magdoff and Tokar 2010). The "Food for Peace Program" (Public Law 480), which was established in 1954, was the first of a series of US legislative acts intended to provide food aid to foreign countries (Winders 2009). This legislation requires much of the funding to be channeled through American firms, however, even if they cannot distribute food at the lowest cost—ADM CEO Dwayne Andreas' official biography states he was one of the main drivers behind food aid laws (Connor 2007). Of the more than $ 1 billion the country spends annually on food aid, approximately two-thirds goes to the ABC firms for purchasing, processing, and shipping these products, frequently at higher than market prices (Provost and Lawrence 2012). This program actually exacerbates hunger by undercutting prices for local food producers and reducing local self-reliance. If the same amount of funds were spent on infrastructure, such as transportation networks or food processing technologies, the receiving nations would be much more self-sufficient (Pearce 2002). Other governments, such as the EU and Canada, have moved entirely to the direct provision of cash as a result (Clapp 2009), and since 2008 the United States has expanded pilot programs that allow up to 25 percent of the international food aid budget to be used for buying local food (Schnepf 2014).

White House safe for a year until the Watergate scandal, which motivated the administration to return the money to Andreas (Krulwich 1996). From the 1990s to until 2012, at least twenty-two Cargill employees served on twenty-one federal advisory committees, including ten committees at the USDA (Sunlight Foundation 2014).

This influence is particularly strong when negotiating trade policies (which would not surprise Johnny Mnemonic). The European Economic Community agreed to exempt oilseeds and meal from tariffs beginning in 1961, for example (Peterson 2009). Numerous Cargill executives have been involved in negotiating multilateral trade agreements since the 1970s. For example, during the Uruguay Round of 1986–1994, Cargill's former vice-president, Dan Amstutz, was employed in the office of the US trade representative; he helped draft the original Agreement on Agriculture implemented by the World Trade Organization, whose provisions allowed the United States and other industrial countries to retain and increase the domestic taxpayer subsidies to commodity farmers (Murphy 2009). Current Cargill executives are among hundreds of corporate "advisors" with access to the draft text of the Trans-Pacific Partnership (TPP), although it is kept secret from US Congress (Schultz 2014).

Milking the system: dairy

Dramatic changes in the dairy processing industry since the 1990s have increased the level of concentration and helped to drive tens of thousands of dairy farms out of business, a transformation that has been described as "milking the system" (Hauter 2012, 211). The key firm involved in this transformation was Dean Foods, which was second in US dairy sales in 2012 (Table 5.2). Dean is the largest processor of fluid milk, although the firm's sales are less than Nestlé, which processes the ingredient into higher-value products like ice cream and frozen pizza. Nestlé is also the largest dairy firm in the world, while Dean is only sixth largest, behind several other European firms and New Zealand's Fonterra (GRAIN 2012).

As noted in the previous chapter, former Dean CEO Gregg Engles has made a fortune consolidating industries, even when the results failed to substantially benefit investors. He moved from the packaged ice business to the dairy

Table 5.2 US dairy sales, 2012

Firm	Sales in billions
Nestlé USA Inc.	$11.174
Dean Foods Co.	$9.320
Saputo Inc.	$7.157
Schreiber Foods	$4.500

Source: Carper (2013).

processing industry at the suggestion of the head of 7-Eleven's dairy operations, Cletes (Tex) Beshears (Martin 2012). They became partners and started with a $100 million leveraged buyout of Suiza in Puerto Rico, followed by thirty-nine additional acquisitions. When they acquired Dean Foods they used its name for the combined firm, after Dean had itself made more than two dozen acquisitions (Hendrickson et al. 2001). Recalling that only big can serve the big, Engles had observed consolidation trends in retailing and positioned the firm to be one of the few that was able to supply Walmart and other national supermarkets. However, Dean continues to maintain at least three dozen regional brands and presents consumers with an illusion of diversity (Figure 5.1).

The CR4 for fluid milk processing by 2002 was 43 percent, doubling in just five years (Shields 2010). Dean's retail market share by itself is even higher in many parts of the United States. It is typically 70 percent in most large metropolitan areas according to the president of the California Farmers Union, and an industry watchdog publication, the *Milkweed*, estimates shares of more than 75 percent in Michigan, Southern New England, Northern New Jersey, and Eastern Pennsylvania (Hardin 2010). The firm's acquisition of Foremost Farms in 2009 went unreported to the federal government, as the sale price was believed to fall under the mandatory limits. When the Department of Justice learned about it the following year, they estimated it resulted in a 57 percent market share in the Upper Midwest and filed an antitrust suit to oppose the deal. Dean fought this effort, and a settlement was reached that allowed the firm to keep one of the two plants it had acquired (Department of Justice 2011a).

This somewhat unusual enforcement effort followed a number of accusations in previous years that Dean had manipulated prices, in collusion with the largest remaining dairy farmer cooperative in the US, Dairy Farmers of America (DFA). These two organizations faced several large, private antitrust lawsuits alleging that DFA management accepted low prices in exchange for being the exclusive supplier to Dean, to the detriment of its farmer members. Dean settled one lawsuit for $30 million in 2010 and another for $140 million in 2011, while DFA agreed to pay $159 million in 2013 and $50 million in 2014. Although both firms admitted no wrongdoing in these cases, they agreed to change some of their business practices as a condition of some agreements (Rosen 2013).

The payouts were reduced by one-third or more to pay attorney fees and then divided among more than 16,000 dairy farmers. The resulting amounts were dramatically lower than payments to some well connected individuals; business partners of DFA's former CEO Gary Hanman and Dean's Engles received $80 to $100 million each, just two years after making investments of $5 to $7 million (Martin 2012). In addition, the low prices that resulted from DFA and Deans'

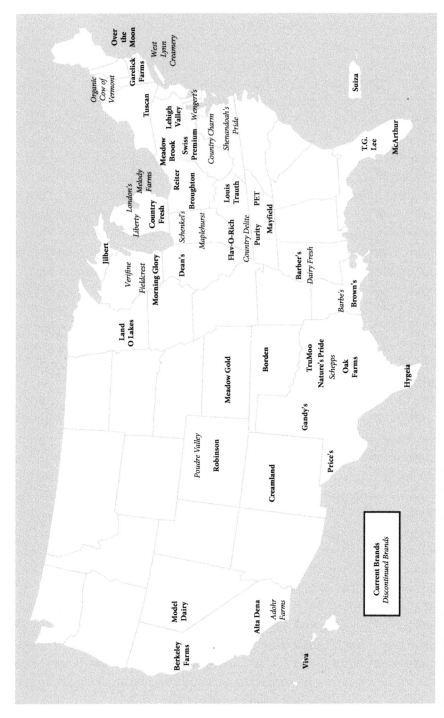

Figure 5.1 Map of brands acquired by Dean Foods.

actions gave farmers far fewer options for selling their milk and thereby contrib-uted to the demise of more than one-third of dairy farms in the states covered by the lawsuits in just a seven-year period (Guebert 2012). The Department of Justice conducted a two-year investigation of anticompetitive practices at the behest of three senators, which resulted in a report recommending specific actions against DFA and Dean but none were taken (Martin 2012).

Paradoxically, DFA's origins can be found in resistance to the same prac-tices implemented by its managers in recent decades (Mooney 2004). In the late 1800s, farmers organized to counter the power of the "milk trust," or buyers in urban areas who colluded to lower the prices they paid to farmers, and by 1909 there were more than 2,700 dairy cooperatives in the United States (Schneiberg, King, and Smith 2008). Although these numbers have declined substantially in recent decades, milk still typically flows through a cooperative before reaching a processor. DFA was formed by a merger of four large cooperatives in 1998, which resulted in more than 15 percent of all US dairy farms coming under its control, as well as a more bureaucratic organi-zation (Gould 2010).

In addition to alleged manipulation of milk prices, DFA has been involved in the manipulation of cheese prices. Hanman bragged about increasing the price of cheese and raising sales revenues by more than 1 billion dollars while head of DFA, which led to federal charges. Two other executives were also implicated, and together they paid $12 million in penalties to the US Commodity Futures Trading Commission, and a $46 million settlement in a subsequent class action lawsuit (Natzke 2013). DFA was able to manipulate the Chicago Mercantile Exchange Cheese Spot Market because it is a "thin" market involving very few buyers and sellers—it accounts for less than 2 percent of US sales (Gould 2010). Nevertheless, it is very influential in determining national prices because there are so few open markets for price discovery. DFA was not the only firm to benefit from this system; in the 1990s, Kraft, Land O'Lakes and Borden faced accusa-tions that they dumped low price cheese on a commodity exchange in order to reduce the prices they paid for dairy inputs (Fulmer 2000).

Another area of price manipulation involves bid rigging of supply contracts for public schools. In the 1990s, executives at Dean Foods, Borden and Pet were accused of meeting to decide which firm would submit the "low bid," which in a typical case was 14 percent higher than if they were actually compet-ing (Henriques 1993). More recently, when its only competitor for school milk contracts went out of business in west Texas, Dean immediately increased prices (Hardin 2010). Public bidding in these cases results in slightly easier detection than manipulation in more private markets.

The dominance of DFA and Dean, and their ability to keep input prices low has encouraged some conventional dairy farmers to convert to organic production. Although Horizon Organic was owned by Dean until 2013, and is a major buyer, its competitors include CROPP Cooperative/Organic Valley and smaller, regional organic farmer cooperatives. With more options for supply contracts, the result is a higher farmer share of the retail price (Maltby 2012). Organic milk has fewer characteristics of a commodity, and more closely resembles a differentiated, niche market product. Some pork producers have similarly changed production practices to escape the grasp of increasingly powerful processors.

Controlling supplies: pork

Pork processors have had an even greater influence on production than dairy processors in recent decades through increasing vertical integration. By becoming more involved in raising hogs themselves and/or increasing the number of contracts with larger-scale pork producers, these firms were able to reduce the number of transactions occurring via public auctions. This control of the majority of supplies resulted in thin markets, and had a dramatic impact on prices—by the late 1990s independent pork producers could only sell hogs for well below their cost of production, as low as 8–10 cents per pound when retail prices averaged $2.20 per pound (Barboza 1998).

If prices on spot markets begin to rise, processors now have the ability to stop buying and switch to their "captive" supplies until the market price falls to a level they find more acceptable. An additional factor in keeping spot markets thin is the practice of processors selling hogs to each other, such as Smithfield's sales to Iowa Beef Processors (IBP) in the late 1990s (Connor et al. 2002). Chuck Wirtz, an independent pork producer from Iowa explained another strategy of large processors in his testimony at joint DOJ-USDA hearings on competition in agriculture: "when they need pigs, they know how to buy them so as not to influence the cost of all their pigs" (DOJ USDA 2010, 204). He stated that these buyers would wait until one minute after the USDAs mandatory price reporting cutoffs, so that the information would be delayed from reaching other sellers as long as possible.

Spot markets accounted for 87 percent of hog sales in 1993, but by 2010 this number had declined to just 5–7 percent; farm numbers were similarly affected during this period, with a loss of approximately 172,000 hog farms, a 73 percent decline (Lawrence 2010). The majority of hogs are now produced under contract with large processors. Increasingly these contracts follow the model developed in poultry, in which the farmers own the land and the buildings, and the processors retain ownership of the livestock, feed and other inputs. Although the technologies for large-scale, confined hog production were developed in the 1970s and 1980s, it was not until banks began providing loans to farmers on the basis of contracts with processors that this system rapidly expanded (Brehm 2005). Critics have raised concerns that livestock producers are being "chickenized," as they take on debt, the production risks and responsibility for the disposal of waste, while processors have the power to change or cancel contracts at any time (DeLind 1995; Ogburn 2011). In 1999, for example, Cargill was charged by the USDA with underpaying $1.8 million to farmers by changing the way it calculated the lean meat content of hog carcasses.

The CR4 for pork processing has increased substantially since 1982, when it was estimated at 36 percent, as leading pork processors in the United States have used mergers and acquisitions to expand both vertically and horizontally (Wise and Trist 2010). The top-ranked pork processor Smithfield (Table 5.3), for instance, had also gained the top ranking in pork production by the end of the 1990s. In that decade, it acquired Carroll's Foods, Murphy Farms, and several other firms, followed with the acquisition of Premium Standard Farms in 2007. Smithfield also purchased Farmland Foods' pork processing plants in 2000 to help maintain its dominance throughout the entire chain.

As the result of more recent ownership changes, half of the top four firms are now headquartered in other countries. Smithfield was acquired by the Chinese firm Shuanghui in 2013. As an illustration of how transnational capitalist investment has become, the Wall Street firm Goldman Sachs had equity in both corporations (Terlep 2013). In 2007, the third-ranked pork processor, Swift, was purchased by JBS of Brazil, which also gave the combined firm the top position globally in beef processing. Brazil's public national development bank took a

Table 5.3 US pork slaughtering capacity, 2013

Firm	Share (percent)
Smithfield-Shuanghui	26
Tyson	17
JBS Swift	11
Cargill	9
	CR4: 63

Source: Tyson Foods (2014).

13 percent holding in JBS that year in exchange for the provision of financing to make acquisitions, which played a significant role in the growth of the firm (Etter and Lyons 2008).

The Packers and Stockyards Act was passed in 1921 in an attempt to limit the power of the meat trust. The processors' buying power resulted in an oligopsony, and concerns about the impact on farmers motivated greater regulation of the industry (Kelley 2003). In recent decades, antitrust lawsuits have resulted in some favorable settlements for dairy farmers, but livestock farmers have had less success. A group of cattle ranchers, for example, brought the first class action suit under the Packers and Stockyard Act against IPB (later acquired by Tyson, in 1998). They accused the firm of manipulating prices through the use of captive supplies, and in 2004, a jury awarded $1.28 billion in damages, the amount estimated to have accrued to the firm as a result. The judge, who was a Reagan appointee, set aside the verdict and even required the ranchers to pay Tyson's court costs, with the rationale that competition in the market as a whole was unharmed (Domina 2004). One of the plaintiffs, Mike Callicrate, said: "It was like a woman losing her purse to a thief having to prove the theft of her purse damaged women with purses everywhere" (2015). The judge's decision was upheld on appeal, which likely influenced regulators refusal to significantly challenge most subsequent acquisitions in meat processing industries (Ward 2010).

The impacts of these changes have affected other groups, in addition to farmers. Meat processing was a heavily unionized industry before the 1980s, but now has much lower wages and unionization rates. Deskilling and speeding up production lines played a role, and contributed to a significant rise in injuries (Compa 2004; Pachirat 2011). Another factor was the closure of unionized plants and moving to "right-to-work" states that are less amenable to labor organizing (Jackson-Smith and Buttel 1998). The composition of the labor force has changed as a result, and most line workers are recent immigrants from Latin America and Southeast Asia. Some of these employees are in the United States illegally, and the fear of being deported contributes to a more docile labor force. In 2006, for example, raids were conducted at six meatpacking plants, and approximately 1,300 workers were accused of immigration violations. One processor, Tyson (Box 5.2), emphasizes its expectation that employees passively accept management's decisions with a sign in both English and Spanish: "Democracies depend on the political participation of its citizens, but not in the workplace" (Striffler 2002, 306).

Box 5.2 "Segment, Concentrate and Dominate": Tyson

Tyson's corporate motto is "segment, concentrate and dominate" (Bonanno and Constance 2010, 132). The firm even filed for US trademark on the phrase in 1998, although it was cancelled the following year. It describes a strategy of first identifying narrow markets in which to compete, then focusing resources on these markets until the firm has gained the top position, followed by identifying other markets to dominate. One successful strategy for segmenting markets is product differentiation, which also helps to increase sales and profits more than is possible with unbranded commodities. Examples include Tyson's *Supreme Tender* brand pork and *Wright Brand* bacon. In 2014, Tyson outbid JBS Swift/Pilgrim's to buy Hillshire Brands, which included Jimmy Dean sausage and Ball Park hot dogs, for $8.5 billion. A condition of US antitrust approval was for Tyson to sell a subsidiary, Heinhold Hog Markets, but it only accounted for a tiny fraction of the firm's revenues (Wohl and Hirst 2014).

Although the corporation has a 17 percent share for second place in pork production, it "dominates" with a leading 21 percent share of chicken production (CR4 54 percent), as well as a leading 26 percent share in beef packing (CR4 85 percent) (Tyson Foods 2014). The firm's origins are in poultry, but it has expanded into other types of meat or "protein" via a number of acquisitions, including a major pork production facility in North Carolina in 1977, and the giant beef processor IBP in 2001 (Wise and Trist 2010). Tyson was an innovator in shifting poultry production risks to its contractors, and is now doing the same with pork producers (Constance and Bonanno 1999). The firm is also expanding geographically, particularly in Latin America and Asia, facilitated by NAFTA and other trade agreements (Constance, Hendrickson, and Howard 2014a).

Tyson Foods was founded by John W. Tyson in the 1930s, just before the poultry industry began its transition to larger scale production. This was the first livestock industry to breed animals for better tolerance of large-scale confinement, and to move much closer to the factory model. In the 1950s, Tyson's son Don convinced him to borrow money from banks in order to expand the market for chicken. The family's willingness to continually reinvest profits in order to increase the firm's power gave them an advantage over competitors in a low margin business (C. Leonard 2014). Tyson Foods was able to survive cyclical downturns better than other firms when markets were more fragmented and remained one of the last firms standing as the industry became an oligopoly.

Processing facilities have also become fewer, larger and more concentrated geographically (Wise and Trist 2010). The decline in the number of smaller facilities has been a key bottleneck for producers that sell in direct markets or to local retailers. Shipping livestock to more distant processing plants raises animal welfare concerns and can be too expensive for small-scale producers (Carlsson, Frykblom, and Lagerkvist 2007). One response has been to develop mobile slaughterhouses that can be brought to the farm, often with the aid of government or nonprofit funding (Gustafson 2012). A small number of pork producers have also formed cooperatives to coordinate access to processing plants, or even own such plants. These tend to focus on smaller, niche markets, with attributes such as pasture-based production, and/or refraining from artificial growth hormones and antibiotics (Grey 2000). Some of these, such as Ozark Mountain Pork Cooperative, struggled to overcome substantial debts before achieving profitability, demonstrating the difficulty of creating alternatives within a system shaped to favor dominant firms (Estabrook 2013).

Transparent markets?

There was a brief window of time, from 2008 to 2011, when activists advocating for greater US government enforcement of antitrust laws grew slightly more optimistic. After decades of effort, the 2008 farm bill included provisions requiring the USDA to write rules for competition in the meat industry. The following year, newly elected president Obama appointed strong antitrust advocates to high-level positions in the Department of Justice and the USDA, which contributed to a jointly organized series of five workshops to examine competition issues in agricultural industries in 2010.

During these years, some concrete actions were taken that hindered consolidation. JBS, for example, proposed to acquire National Beef, but the deal fell apart after it was opposed by DOJ and several states. Notably, this would have reduced the number of firms controlling more than 80 percent of the market from four to just three—the current CR4 is already higher than when the meat trust was broken up in 1920s. In addition, as noted above, Dean Foods' attempt to acquire a relatively small processor, Foremost Farms, was partially blocked during this time.

Yet by January 2012, the most promising appointees had all resigned. Christine Varney, for example, the assistant attorney general who told one poultry farmer at a DOJ-USDA workshops that if he experienced any retaliation to give her a call, left for private practice the following year (Andrews 2012). J. Dudley Butler, previously a lawyer who had represented poultry contractors, was tasked with running the Grain Inspection, Packers and Stockyards Administration (GIPSA) at the USDA and writing the competition rules for the meat industry. He was quickly thwarted by the power of their lobbyists, as members of Congress have kept the program from being funded since rules were proposed in 2010, via riders to annual appropriations bills (National Sustainable Agriculture Coalition 2014).

Regulations at the state level have also been dismantled or suffered from lack of enforcement. Anti-corporate farming laws were passed by nine states in the late 1970s and early 1980s, which prevented corporations from owning land or livestock (Welsh, Carpentier, and Hubbell 2001; Lobao and Stofferahn 2008). Most of these have been removed by legislators or court decisions, or are no longer enforced; the few that remain, such as Nebraska's ban on livestock ownership, are under heavy pressure to be dismantled (Gerlock 2014). Anti-price gouging laws are another protection for farmers in theory, such as New York State's regulation that restricts the price of milk to no more than 200 percent of costs. A New York City Council survey, however, found that 86 percent of retailers were in violation of this statute (Lewis 2008).

Initiatives to create more transparent pricing as a voluntary option for consumers have experienced somewhat more success. Fair Trade is the most prominent example, it provides participating farmers with a minimum price (Renard 2003). The scheme also seeks to reduce the number of network connections between producers and consumers, and includes additional social and environmental criteria (Jaffee 2007). Although it only applies to less industrialized nations, there are efforts to create similar programs in industrialized nations (Howard and Allen 2008), such as the Food Justice Certified label in the United States and Fair Deal Food in the UK. The latter has worked with the supermarket chain, Booths, to replace its private label milk with a "Fair Milk," which claims to provide farmers with the highest market price in the nation (Shackleton 2014).

This chapter explored the increasing power of commodity processors and their ability to control markets both upstream and downstream from this stage of the food system. Campaigns to reign in this power via government actions have achieved few victories, but some tiny alternatives are demonstrating the potential to create more direct and transparent processing markets. The following chapter takes a closer look at the upstream segment, farming, and ranching, which has long suffered from many of the negative impacts of dominant processors' power. Although farming and ranching is the least concentrated stage of the food system, it is also experiencing changes that are increasing rates of consolidation, resulting in even fewer people producing our food.

Chapter 6

Subsidizing the treadmill: farming and ranching

This isn't a business. I've always thought of it more of a
source of cheap labor, like a family.
—Prof. Farnsworth (*Futurama*)

How have government policies subsidized farming, and who really benefits from these subsidies? This chapter explores these questions, with a focus on soybeans, milk, pork, and leafy greens. For each of these products, taxpayer funds have reinforced the advantages of larger-scale operations at the expense of smaller operations. The greatest benefits, however, have accrued to the dominant firms that are: (1) located upstream and sell products to farmers, (2) located downstream and buy farm commodities, or (3) have vertically integrated into numerous stages of the food system, including production. This has accelerated farm consolidation and reduced the power of remaining farmers relative to these firms.

Agriculture is the least concentrated stage of the food system, and social scientists have been asking why for more than 100 years. Karl Kautsky (1988) sought to understand the persistence of smaller farms in his book *The Agrarian Question*, first published in 1899. He noted the tendency for farmers to exploit themselves (and as Prof. Farnsworth pointed out, their family members) by working harder and for longer hours, when compared to other laborers, in their desire to remain on their own land. Since Kautsky's era, a number of social scientists have explored other explanations that contribute to farm survival, such as barriers or obstacles to capital. These include the extensive amounts of land required to grow crops or raise animals, the long production times that are dependent on biological processes, as well as factors such as weather, pests, or diseases that can have catastrophic impacts on yields (Mann and Dickinson 1978; Lewontin and Berlan 1986). In other words, the factory model developed by many other industries does not always work well when applied to agriculture.

One response has been to industrialize off-farm processes, which Goodman, Sorj, and Wilkinson describe as *substitutionism* and *appropriationism* (1987). Substitutionism applies to downstream commodities, such as replacing butter with a more industrial counterpart, margarine. Appropriationism refers to under-mining on-farm processes via upstream inputs that can be produced in factories, such as fertilizers instead of manures, and machines or pesticides instead of manual labor. Changing social relations, such as one-sided contracts that give capitalists a high degree of control over farm processes, even without direct ownership, are another factor in reducing farmer self-reliance and deepening processes of exploitation (Watts 1994).

These processes have contributed to a decline in the number of farmers in the United States, from 6.8 million in 1935 to less than 2 million by 1997, and farming continues to become even more concentrated (Marion and MacDonald 2013). Mid-scale farms and ranches make up the majority of agricultural oper-ations, but these are experiencing the fastest losses, as large operations are typically becoming even larger (Guptill and Welsh 2014). Less than 100,000 farms accounted for approximately 66 percent of agricultural sales in 2012, for example (USDA 2014a). At the other end of the spectrum, smaller-scale farms and ranches are increasing in number, although they account for just a minus-cule percentage of all sales.

Mid-scale farms have been described as being on a treadmill, because demand for agricultural products is often inelastic, and producing more has the effect of reducing prices. Successful efforts to increase farm output in these cases results in the same or lower incomes, which further reinforces this cycle (Cochrane 1979). Because farmers are usually not organized enough to restrict supplies, many have no choice but to buy more technologies, inputs and/or land in their attempt to maintain their economic viability. This leads to spending increasing amounts of money to simply obtain the same net income; much of this additional money flows to dominant capitalists, such as input suppliers. Larger-scale farms and early adopters of new technologies tend to stay slightly ahead In the process. Smaller and mid-scale conventional farmers "fall off" the treadmill and their land ends up being "cannibalized" by larger farmers (Cochrane 1979, 431). The technology treadmill describes a long-term trend, but in the short term, prices can be more cyclical. This contributes to even greater instability, because when prices rise, production typically increases even more.

This treadmill is heavily subsidized by the government, which encourages overproduction and supplements the resulting low prices. Direct subsidies glob-ally amount to approximately 1 billion dollars per day, the majority provided by the governments of the United States, EU, and Japan (Wise 2004). The United States has spent approximately $10 billion annually in recent years. These payments are not distributed equally but are disproportionately weighted to the largest farms—of the top twenty US agricultural subsidy programs, for example, 85.5 percent of funding goes to the top 20 percent of recipients (Figure 6.1). For only the smallest of these programs, sugar beets, did the top 20 percent receive less than two-thirds of the payments.

The distribution of subsidies played an important role in the disproportionate loss of minority farmers, as more affluent farmers are of predominantly European heritage (USDA 2014a). In addition, minority and women farmers faced discrimination from USDA employees when applying for government programs, such as direct discouragement, delays in processing paperwork and high rejection rates. A class action lawsuit filed by black farmers resulted in a settlement of $2.25 billion dollars, and a settlement with Native American farmers totaled $760 million. Similar suits by Hispanic and women farmers motivated the USDA to provide $1.3 billion in additional funds for these claimants (Doyle 2013).

Indirect government subsidies to farming and ranching, which tend to be underestimated even by politically conservative analysts (e.g., Riedl 2007), also accelerate the treadmill. These may include: (1) funding for research and development to increase production, (2) university extension services to promote the adoption of new technologies, (3) purchasing excess production, (4) subsidies for inputs such as fertilizers, fuel or irrigation, (5) government enforced "checkoffs" that require producers to pay for marketing, (6) regulations that enable the costs of pollution or soil loss to be externalized, and (7) regulations that increase barriers for competitors, such as import tariffs.

Both direct and indirect subsidies receive political support on the basis of the agrarian ideal, which valorizes independent, self-reliant farmers as upholders of democracy. This view was famously espoused by Thomas Jefferson but has its roots in much older political philosophies; agrarians believe farmers' strong ties to the land motivate them to effectively resolve tensions between community and self-reliance (Thompson 2007). Social science research has since investigated the potential linkages between farming structure and positive community outcomes. In the 1940s, Walter Goldschmidt reported more indicators of well-being in a community characterized by numerous small farms, when compared to a community composed of fewer large farms (Goldschmidt 1978). Subsequent research has largely supported these findings but continues to explore the reasons behind them, such as the relative importance of scale, social class and geography (Lobao and Stofferahn 2008). Ironically, although

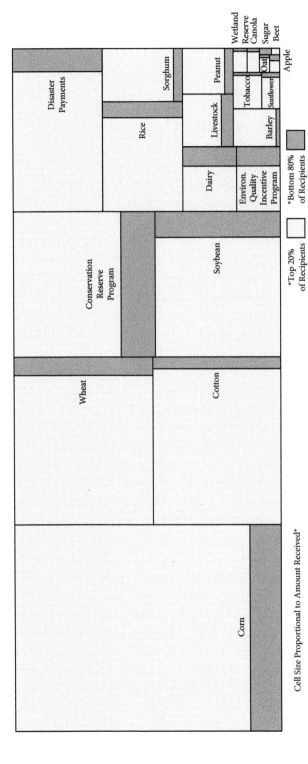

Figure 6.1 Distribution of the top 20 US agricultural subsidy programs, 1995–2012. Data Source: Environmental Working Group (2013).

the rhetoric of "family farms" is frequently used to justify subsidies, the majority of such operations receive little benefit from the current system (Goodwin, Mishra, and Ortalo-Magné 2011).

Pricy premiums? Soybeans

Soybean farming in the United States remains dominated by mid-size growers. Because soybean and corn farmers employ similar practices and equipment, many farmers grow both, often in rotation. Together, they account for more than half of land planted in crops nationally. Production of soybeans and corn has become more capital-intensive than labor-intensive, and as a result, the majority of farmers engage in this occupation part-time. As Kautsky observed, farmers are willing to exploit themselves to remain living on farms, and more recently this involves a reliance on off-farm employment income to keep many operations economically viable; more than 90 percent of all US farms have at least one family member working at off-farm jobs (Brown and Weber 2013). In addition, most would lose money on their agricultural operations without government subsidies.

Direct subsidies have been a part of US agricultural policy since the 1930s. While the details are constantly changing, since that time they have remained focused on just a handful of commodities, which, not coincidentally, continue to account for most of crop production: corn, wheat, soybeans, cotton, and rice (Peterson 2009). Initially, the federal government established price floors for wheat and cotton, and if prices fell too low, directly purchased these crops to store and sell at a later time. These incentives quickly resulted in overproduction, which then led to programs that paid farmers to keep land out of production and keep supplies lower. In subsequent decades, subsidies expanded to other commodities and payments tended to increase. By 1996, pressure to reform farm subsidies resulted in legislation that proponents claimed would phase them out but actually resulted in increasing these expenditures—the withdrawal of planting restrictions resulted in overproduction and low prices, followed by "emergency" payments (Starmer, Witteman, and Wise 2006).

These subsidies have resulted in much lower input costs for grain processors and meat producers, as more than 80 percent of soybeans are used for animal feed. Most of the global trade flows from producers in the United States and Latin America to the EU, which produces very little soybeans, and to China, which does not produce enough crops to feed all of its livestock (U.S. Soybean Export Council 2011). An increasing amount of soybeans and corn are also being diverted for use as biofuels (Weis 2010). This outlet benefits grain and oilseed processors, although it also contributes to conflicts with other food industry firms, due to their opposition to paying more for inputs based on these two crops. In response, a number of retailers, packaged food firms and meat processors have formed a "Food Before Fuel" campaign to reform US government subsidies that encourage ethanol production (Baines 2014b).

Soybean subsidies have averaged approximately $1.5 billion per year from 1995 to 2012 (Environmental Working Group 2013). Legislation passed in 2014 ended fixed payments tied directly to acreage, and placed more emphasis on subsidizing crop insurance premiums. This program pays an average of 62 percent of farmers' premiums, which results in the purchase of much more coverage than is needed—insurance companies received $1.3 billion for all commodities through this channel in 2011, and lobbied heavily to maintain it (Dayen 2014). In addition, the legislation covers "shallow losses," meaning it subsidizes insurance for market price declines that are less than the insurance deductible. In other words, if crop prices go down, payments go up. Even strong advocates of farm subsidies have criticized these programs (Ray and Schaffer 2012), and some analysts suggest that the total payments could easily exceed those made in previous years (Smith, Babcock, and Goodwin 2012; Sewell 2014). In addition, because the payment limits are quite high, up to $250,000 per farm, most of these subsidies would continue to go to the largest farms.

Some USDA economists suggest that subsidies have likely played a role in accelerating farm concentration and the loss of mid-size farms, such as by increasing the advantages of larger, more profitable farms and reducing their risks (Key and Roberts 2007). In the case of soybeans, the median farm size doubled between 1987 and 2007, from 243 to 490 acres (MacDonald, Korb, and Hoppe 2013). The amount of land planted in this crop has increased dramatically in recent years, helped by higher prices, to reach a record 85 million acres in 2014 (Pitt 2014). This has raised concerns that marginal lands are being brought into production, which may have negative impacts on the soil, and lead to inflated land prices as well (DeVore 2012).

Ordering overproduction: Milk

Subsidies that contribute to record production levels are also a challenge facing US milk producers. Price supports and government purchases beginning in the 1930s have resulted in overproduction ever since. From 2004 to 2013, for example, production quantities rose in all but one year, for a total increase of 18 percent (USDA 2014c). During the same period, per capita dairy consumption leveled off, with steep declines in fluid milk consumption. This oversupply results in very low market prices, which primarily benefits dairy processors like Dean and Nestlé.

Direct subsidies for dairy farmers have included marketing orders, price supports, and income loss contracts, although the latter was replaced with insurance subsidies in 2014. Milk marketing orders mandate minimum prices that processors must pay, which differ depending on whether it is sold as fluid milk, or various categories of further processed dairy products. These orders apply to approximately two-thirds of fluid milk; the exceptions are states, such as California, that have developed their own equivalent programs. Milk marketing orders are typically opposed by dairy processors (Shields 2009), but these programs have withstood efforts to dismantle them thus far.

DFA, Dean, and other dominant milk processors have lobbied heavily to maintain price supports and income loss contracts, however, which have lowered and stabilized the prices they pay. The Dairy Product Price Support System, for example, purchased nonfat dry milk, butter, and cheese at specific prices when market prices fell below this level. The government paid storage costs of approximately $20 million per year for these products, which at one point weighed over a billion pounds. The formula for determining the level of this safety net was frequently subject to the influence of dairy farmers and processors (Belongia 1984). In 1971, for instance, three dairy cooperatives were accused of funneling hundreds of thousands of dollars to Richard Nixon's re-election campaign in exchange for raising price support levels (Lardner 1974). This program was ended in 2014; the government may continue to purchase excess supplies for donations for low income groups but only when prices are low.

Milk income loss contracts (MILC) were established in 2002. This was a counter-cyclical program, meaning that payments were only triggered when market prices dropped below a regional monthly minimum, using a formula that incorporated the costs of feed. It was limited to the production equivalent of an operation with about 160 cows and therefore not popular with larger producers (Shields 2009). This program was eliminated by legislative changes in 2014, but another program was added, which, similar to field crops, subsidizes insurance premiums for payouts when prices drop below a given threshold. Although the details differ, the impacts are expected to be comparable to the MILC program, except that there are no longer limits on payments (Bozic et al. 2014).

Critics suggest that despite these modifications, the programs will continue to encourage higher production and thereby increase profits for processors and manufacturers (Collins 2014). This explains why dominant firms have continued to support these Byzantine subsidies. The dairy farmers who are supposed to be the target of these programs tend to be caught in a price-cost squeeze, receiving lower returns for their products while paying increasing prices for inputs. Retail prices for consumers, in contrast, tend to be "sticky," continuing to increase when costs rise for powerful processors and retailers but less likely to fall if these costs decrease (Shields 2010, 5).

Indirect government subsidies have also accelerated the treadmill for dairy farmers. Funding for technologies to raise production per cow have been very effective, resulting in an average increase of six to seven times as much milk when compared to a century ago (Kurlansky 2014). Some of these technologies include selective breeding, artificial insemination, antibiotics, and growth hormones. When genetically engineered, or recombinant, bovine growth hormone (rBGH) was developed, for example, it increased milk production by an average of 10 percent. Trials were conducted at public universities, including Cornell University and the University of Vermont—just a few years after the government paid farmers to slaughter more than a million dairy cows in order to reduce production levels. Operations have become more industrial as a result of automated milking parlors and other technologies subsidized by government research. This has led to dairies of 30,000 or more cows, such as Threemile Canyon Farms in Oregon and Fair Oaks Farms in Indiana (Estabrook 2010).

As factory scale operations have increased, particularly in arid regions in the Western US, the number of small dairy farms has decreased—although this restructuring has not been quite as dramatic as the pork industry in recent decades (Jackson-Smith and Buttel 1998). The number of dairy farms declined from 202,000 in 1987 to 70,000 in 2007, during which time the average herd size increased from less than 50 to approximately 131 cows (Gould 2010). The rise in large, highly capitalized dairy farms has made it more difficult for the industry to respond to declining demand; these dairies are less likely to cease operations than smaller farms with less than 80 cows, which may simply shift their emphasis to other crops or livestock when prices are low (Vander Dussen 2008). More than a dozen state governments have provided additional support for this restructuring; during the financial crisis of 2008, they recruited wealthy farmers from Holland to build large-scale confinement dairy operations, touting their cheaper land and less restrictive regulations compared to the EU. Some of the families that agreed to immigrate went bankrupt soon after investing millions of dollars, due to higher than expected feed costs and lower milk prices (Etter 2010).

In contrast to these efforts to increase supplies, some regional dairy cooperatives have worked to restrict the amount of milk produced by their members, in order to help keep prices higher (Vander Dussen 2008); this is allowed by the Capper-Volstead Act of 1922, which gave cooperatives limited exemptions from antitrust legislation. The alternative of organic milk is discussed in more detail in Chapter 8, but pastured or grass-fed milk is another rapidly growing niche market that garners price premiums for dairy farmers. Snowville Creamery, for example, has expanded from sourcing from one farm in Southern Ohio in 2007, to include a dozen nearby farms seven years later (Taylor 2014). Other dairy farms are bypassing processors entirely by selling raw (unpasteurized) milk directly to the public, although government regulations have made this difficult in most states (Gumpert 2009).

Paying for their own demise: Pork

Every farmer who sells hogs is required to pay a "checkoff" of 40 cents for every $100 in sales, which goes to the National Pork Board. This quasi-governmental organization was established in 1985 to conduct advertising, research, and education on behalf of the pork industry, but it is prohibited from lobbying; similar programs fund campaigns for other commodities, including "Got Milk?" and "Beef. It's What's for Dinner." Smaller-scale hog producers have criticized the Pork Board for handing over much of the approximately $50 million collected per year in the form of contracts with the National Pork Producers Council (NPPC), which typically promotes the interests of larger-scale producers and processors. The NPPC, for example, used more than $50,000 of checkoff funds to hire a consulting firm to spy on family farm organizations in 1996. The targeted groups, Iowa Citizens for Community Improvement, Missouri Rural Crisis Center and the Land Stewardship Project forced the NPPC to pay back the funds and triggered an investigation that led the USDA to require greater separation between the two organizations (Oates 2000).

In 2006, however, the National Pork Board agreed to pay the NPPC $60 million over twenty years for the intellectual property rights to the marketing slogan, "Pork: the Other White Meat." This slogan was mostly retired just four years later and replaced with "Be Inspired." The Humane Society of the United States filed a lawsuit, alleging that this was a thinly disguised payment for NPPCs lobbying efforts, but it was dismissed by a federal court in 2013. Documents related to the sale suggest there was little justification for such a high price, as no other group would be likely to use the slogan (Wilde 2013b).

Groups of producers have attempted to overturn these checkoffs, through both votes and lawsuits. Although pork producers voted to terminate the check-off in 2000, the following year it was determined to be merely an "advisory" vote by the incoming Bush appointee for Secretary of Agriculture, Ann Veneman, and the program was kept in place. In 2001, the US Supreme Court ruled in favor of mushroom growers, agreeing that the checkoff program violated their free speech rights—this was followed by two similar decisions for pork producers in lower courts (Farm and Dairy 2002). In 2005, however, the US Supreme Court ruled that a beef producers' checkoff was "government speech." Promoting increased meat consumption as a government goal would seem to be contradictory to its goal of improving public health (Wilde 2013a), but the decision effectively kept other checkoffs, including pork, fully in place.

Checkoff funds have been allocated for more than 100 studies on improving large-scale pork production facilities (National Pork Board 2014), which contributes to increasing the advantages of these operations. The rise of what the Environmental Protection Agency calls Confined Animal Feeding Operations (CAFOs), or what critics call factory farms, has dramatically transformed the industry. In the mid-1970s, hogs were commonly raised along with other crops, and there were well over half a million hog farmers in the United States (Grey 2000). Although the number of hogs has remained steady between 1992 and 2009, the number of producers declined 70 percent, from over 240,000 to less than 72,000 (McBride and Key 2013). The farmers' share of retail price fell from 45 percent to less than 20 percent during this time (Plain 2003).

As described in the last chapter, the largest pork producer in the United States is also the world's leading processor, Smithfield/Shuanghui (Table 6.1). This Chinese-owned firm also contracts with more than 2,000 hog farmers. Christopher Leonard's (2014) book, *The Meat Racket*, details the impacts of these contracts on producers, such as borrowing half a million dollars to build hog barns, only to have the processing firm drastically reduce the length of the agreement and increase the production requirements. The majority of pork processors employ a tournament system, which forces farmers to compete for the highest production efficiencies, with pay tied to their rankings. Because farmers have no control over the selection of pigs, feed, and other key variables, this tournament is more like a "lottery" (Domina and Taylor 2009, 3).

Table 6.1 US pork production, 2013

Firm	Heads
Smithfield/Shuanghui	868,000
Triumph Foods	381,500
Seaboard Foods	217,000
The Maschhoffs	208,000

Source: Freese (2013).

Pork producers receive fewer direct subsidies when compared to soybean and dairy producers, but indirect subsidies include below cost corn and soybeans for feed, and weak environmental regulations (Starmer and Wise 2007). CAFOs have caused numerous air, water, and land pollution problems, which contributed to a moratorium in North Carolina in 1997, just a few years after making enormous efforts to attract these operations to the state. A government program to provide direct subsidies to mitigate these issues, called the Environmental Quality Incentives Program (EQIP), pays approximately $100 million annually to operators of livestock CAFOs (Gurian-Sherman 2008). The program was originally restricted to smaller operators, but caps were eliminated in 2002 and this resulted in far more disproportionate payouts, up to $450,000 for some operations. Producers in the EU also receive direct subsidies, enabling Smithfield (Box 6.1) to collect millions of euros from these programs while expanding its operations in Poland and Romania and driving smaller farms out of business (Carvajal and Castle 2009).

Box 6.1 Skirting Regulations: Smithfield

Smithfield has a long history of finding creative ways to avoid regulations that are barriers to increasing power. When the firm was in the process of acquiring Murphy Farms' operations they ran into a problem in Iowa—Smithfield was vertically integrated into processing, and Iowa's anti-corporate farming statute prohibited processors from involvement in production. Murphy Farms "sold" its assets to a former manager, Randall Stoecker, giving him a loan of $79 million to pay for them, which he promised to pay back in ten years. Smithfield received the promissory note when it acquired the firm. Smithfield's next move was to file a lawsuit against the state to be exempted from enforcement of the anti-corporate farming law. This was successfully achieved with a settlement, and one condition was that the firm to pay more than a million dollars to Iowa State University for training and scholarships (Food and Water Watch 2008).

Other states have also proved amenable to changing regulations in the firm's favor. Political candidates in North Carolina received more than $1 million from Smithfield's political action committee, which was followed by granting CAFOs exemption from zoning laws and most liabilities from pollution (Carvajal and Castle 2009). In Missouri, the firm threatened to leave the state after a jury awarded $11.5 million in damages to neighbors of a subsidiary's hog CAFO; the state responded by changing the law to limit the amount of damages (Huber 2014).

Smithfield successfully challenged laws while expanding into Poland as well, just before that nation joined the EU. Although the firm acquired a controlling interest of the former state owned processor, Animex, with a hostile takeover, a regulation prevented foreign companies from buying additional land for hog production. Smithfield' solution was to create a front company, Prima Foods, and send its Polish employees out to buy more farms (Public Citizen 2004). That same year, in 2001, the firm was suspected of stealthily changing the nation's definition of liquid animal manure from sewage to fertilizer, thus legalizing its application to land and facilitating much more intensive operations of thousands of hogs. One village granted its approval for the firm to operate there, after $ 4,000 was paid to the mayor's wife to conduct an impact assessment (Food and Water Watch 2008). Some regulations were simply ignored, however. An agricultural ministry investigation in 2003, for example, found that all sixteen Smithfield operations were in violation of various laws, but the small fines that resulted were insignificant to the global firm (Kennedy 2003). Similar expansion efforts in Romania coincided with the loss of 90 percent of hog farmers—affecting more than 400,000 farms—in just four years (Carvajal and Castle 2009).

Shipping water from the desert: leafy greens

The biggest processors of leafy greens, such as Dole and Chiquita/Fresh Express are directly involved in production but also contract with other large lettuce and spinach growers, who may farm tens of thousands of acres. The vast majority of operations are concentrated in California and Arizona but also include Mexico and Canada. Leafy greens are considered "specialty crops" by the USDA and receive far less direct subsidies than commodity crops. The industry did receive $25 million in direct subsidies due to market losses following 2006 *E. coli* epidemic (see Box 6.2), but growers typically benefit from more indirect subsidies, such as those that reduce the costs of water and labor.

Although agriculture in California accounts for just 2 percent of its economy, it consumes 80 percent of the water diverted for human use (Mount, Freeman, and Lund 2014). Much of the leafy green production is in the Salinas Valley, a near-desert, which necessitates irrigation to grow these crops. The irrigation system was established in the early 1900s with the rhetoric of supporting family farmers but is now weighted to the interests of the largest growers. California's water is supplied at rates below full cost, at a savings of hundreds of millions of

---

Box 6.2 Creating Regulatory Barriers: Western Growers Association

The 2006 epidemic of *E. coli* O157:H7 in bagged spinach led to hundreds of illnesses across twenty-six states. It also resulted in three deaths, an expensive recall, and a loss of sales estimated at more than $100 million in the following months. The firm implicated was one of the largest processors in the industry, Natural Selection Foods, which contracted with hundreds of growers. The response of industry associations—such as Western Growers Association, which primarily represent larger growers—was to develop a voluntary set of regulations that was expected to be the model for eventual mandatory government regulations. These regulations were also anticipated to be difficult to meet for more diverse operations, such as Community Supported Agriculture farms. In industry publications, representatives of Western Growers publicly stated that the costly regulations would likely drive smaller farmers out of business, and that this "Darwinian" process would be a positive outcome (DeLind and Howard 2008, 311). The subsequent Food Safety Modernization Act of 2010 required mandatory rules to be developed for leafy greens, and Western Growers lobbied to oppose exemptions for smaller farms, as well as exemptions for specific greens, such as kale (Linden 2013).

The strategy of creating regulatory barriers that primarily benefit powerful firms has been used for more than 100 years. Milk pasteurization laws in many large US cities beginning in 1914, for example, contributed to the decline of small milk distributors and the rise of two national dairy companies, Borden's and Sealtest (Levenstein 1988). The Meat Inspection Act of 1906 was backed by the meat trust because the regulations already applied to their exported products and were expected to create a greater burden for smaller firms. The only real battle was over who would pay for inspection costs, and the large packers succeeded in placing the burden on taxpayers rather than the industry itself (Kolko 1963). By encouraging mandatory regulations, dominant firms essentially create a cartel—they avoid competing to have the best food safety practices and gain the advantage of government enforcement, so that there is no possibility of defection from the standard (Carson 2007). It also allows them to shift responsibility to the government if there is an outbreak of illness or other safety problem, as they can claim they were following the most stringent regulations.

---

dollars per year. More than two-thirds of the savings go to the top 10 percent of farms, at approximately $350,000 per farm, although one operation received $4.2 million worth water one year (Bacher 2004). The USDA EQIP program also provides cost sharing for irrigation equipment, and nationally nearly a billion dollars, or one-quarter of the budget, has gone to this subsidy; this tends to reinforce the problem by encouraging growers to water more frequently and/ or switch to more water intensive crops (Cox 2013). Leafy greens growers also receive significant electricity subsidies to help power their irrigation systems (Sharp and Walker 2007).

Beyond the immediate financial costs of subsidizing irrigation are the ecological costs of "embedded" or "virtual" water. The amount of water "hidden" in food production is receiving increasing attention due to declining water tables and frequent droughts (Rosegrant, Ringler, and Zhu 2009). Heavily irrigated leafy greens from California are transported across the United States, as well as to several other countries, including many areas with more abundant water for agriculture. The impacts of irrigation on ecosystems, including pollution and the loss of endangered species (Postel 2000), highlight the irrationality of these flows when considering the externalized costs.

Although publicly subsidized research helped mechanize and increase the scale of leafy greens production (Friedland, Barton, and Thomas 1981), the industry is still very labor intensive, and extremely dependent on immigrant workers. Approximately half of farm workers in the United States are considered "unauthorized" (Hertz and Zahniser 2013). Although there are reports of frequent abuses, such as underpaying workers and safety violations, the threat of being fired or deported keeps many workers from being willing to speak up (Neuburger 2013). Farmworkers are specifically exempted from many labor laws, but the lack of enforcement for existing statutes allows growers to spend even less on labor costs; farmworkers report average incomes of just $7,500 to $18,000 annually (Melcarek and Allen 2013). In addition, other public subsidies may partially make up for these low wages—such as those for food assistance, health care, and housing—although only a small minority of migrant workers make use of these programs (Taylor and Martin 1997).

Ending direct subsidies?

Every five years in the United States there are debates over the "Farm Bill," the national legislation that determines agriculture and food policy. There are constant calls to end direct farm subsidies but little concrete action. The most recent legislation slashed food stamps instead. In 2013, fourteen members of Congress voted to increase farm subsidies by $9 billion, while they themselves received these subsidies. All but one also voted to cut $20 billion from the budget for food aid to help cover the increase—it was reduced to $8.7 billion over ten years in the final bill. Representative Stephen Fincher of Tennessee, who accepted nearly $3.5 million in farm subsidies said, "the role of citizens, of Christians, of humanity, is to take care of each other, but not for Washington to steal money from those in the country and give it to others in the country" (Shen 2013). The final 2014 bill eliminates payments that are based on historical production levels but retains other subsidies, such as those for insurance premiums, as described above.

The European Union's policy has been to largely "decouple" subsidies from production since 2005, but it pays even more than the United States to farmers for programs such as environmental protection, food safety and quality, and rural development. Approximately € 50 billion, or 40 percent of the entire EU budget, is spent on these subsidies. As in the United States, just 20 percent of farmers receive approximately 80 percent of the payments, but more EU funds go to large processors, including those headquartered elsewhere (Waterfield 2009). The names and amounts paid to more than 90 percent of the recipients—likely including the Queen of England—are kept secret (BBC 2012). Recent changes have placed more emphasis on crop diversification and setting aside portions of land for permanent pasture and conservation, which is supposed to result in a more equitable distribution of payments (Wisdorff 2013).

Some critics have suggested that eliminating direct subsidies in industrialized countries would reduce the negative health impacts of cheap access to soybeans and corn (Elinder 2005). Other analysts, notably University of Tennessee's Darryl Ray and colleagues, suggest that this strategy would have little influence on production levels, and instead argue for reforms that would take more land out of production and create stronger price floors (Ray, Ugarte, and Tiller 2003). Although this proposal focuses on traditional commodities, it is likely compatible with suggestions to shift more subsidies toward healthier foods, such as fruits and vegetables (Franck, Grandi, and Eisenberg 2013). Specialty crop subsidies received extremely modest support in the 2014 legislation, but with more than $100 million a year spent on lobbying by "agribusiness" in the previous year, substantial reforms appear unlikely in the near future (Nestle 2014).

New Zealand is an example of an industrialized country that largely eliminated direct subsidies, which resulted in major farmer protests in the mid-1980s. Although social scientists do not agree on its impacts, most accounts suggest that, after a period in which farmers increased their self-exploitation while adjusting to the dramatic changes, farm losses have been relatively low (Cloke 1996; Smith and Montgomery 2004). The Cairns Group, which is composed of New Zealand and eighteen other export dependent countries, exerts pressure on other nations to reduce their tariffs and agricultural subsidies. The inability to reach an agreement on this issue during multilateral trade negotiations contributed to the failure of the Doha Round of World Trade Organization meeting in 2008, which allowed contentious subsidies to continue (Otero, Pechlaner, and Gürcan 2013).

This chapter has explored how farm subsidies speed up the treadmill and lead to the continued loss of farms, particularly mid-size farms. It also detailed how buyers of farm commodities and sellers of farm inputs exploit these payments, which explains why many of these interests have lobbied to maintain the current system. The following chapter examines the input suppliers in more detail, and how farmers are becoming even more dependent on these firms as they find ways to privatize formerly common resources.

Chapter 7

Enforcing the new enclosures: agricultural inputs

Corporate society takes care of everything. And all it asks of anyone, all it's ever asked of anyone ever, is not to interfere with management decisions.
—Bartholomew (*Rollerball*)

Between the sixteenth century and the General Enclosure Acts of the nineteenth century, large landowners consolidated their holdings and enclosed much of the common lands in England. Small farmers, cottagers and squatters were driven from lands they previously inhabited, and others lost their customary rights to graze animals and gather fuel from these areas. As a result, many people were forced to migrate to cities, where they provided a source of cheap labor for capitalist factory owners (Wood 1998). This protracted process required tremendous government intervention, which was justified by capitalists at the time—despite strongly conflicting with their legitimizing ideology of laissez-faire, or non-interference with the functioning of markets (Perelman 2000).

In the last few decades, a similar process has affected the inputs used by farmers and ranchers around the globe (Kneen 1993). Seeds and animal breeds have been an open access, common resource for millennia, developed and improved through the efforts of countless generations of people. Increasingly, however, dominant firms are appropriating these resources. Two primary means have been used to privatize seeds and breeds and make them more amenable to capitalist strategies of accumulation, (1) government-enforced intellectual property protections have facilitated legal enclosures of living organisms and (2) government-subsidized technological innovations are enabling biological enclosures (Kloppenburg 2004). This increased ability to exclude traditional practices has triggered a wave of consolidation in the seed and animal genetics industries (Howard 2009c). With their growing size and power, dominant firms are pushing the boundaries of intellectual property rights established by patents on genetically engineered (GE) organisms to restrict open access to other kinds of seeds and livestock, such as hybrids.

This chapter focuses on the seed-chemical complex and livestock breeding industries, and the role of governments in enforcing legal protections they have been granted—such protections are critical for implementing processes of appropriation described in the last chapter. It also details some of the negative impacts of these trends, including the potential to threaten the future of the food supply, due to the loss of biological diversity and local knowledge. It then describes efforts to counter these trends and increase self-reliance, particularly by challenging intellectual property protections.

Criminalizing self-reliance: the seed-chemical complex

After the Second World War, agricultural chemical companies in the United States were very successful in their efforts to reduce the self-reliance of farmers and increase the use of synthetic insecticides and herbicides (Lewontin and Berlan 1986). By the 1980s, however, these companies had difficulty exceeding the growth rates of other sectors of the economy (Moretti 2006). The vast majority of farmers that remained had already adopted their proprietary (brand name) pesticides, and growing public concern about pollution discouraged higher rates of application, leaving little room for rapid sales growth (Lewontin 2000). This crisis led firms to increase their marketing efforts in less industrialized countries, but there they encountered the problem of lower purchasing power, and a strong preference for less profitable generic pesticides (Fernandez-Cornejo and Just 2007).

One response was to buy increased market share through mergers and acquisitions. Thirty firms in the 1970s had combined into just six firms by 2001, and the CR4 for global pesticide sales increased from 28 percent in 1994 to 53 percent in 2009 (Moretti 2006; Fuglie et al. 2011). Critics dub the resulting market structure the "Big Six," and the numerous alliances between them are described as a cartel (Shand 2012). Yet this strategy also failed to provide a long-term solution to the problem of how to outpace the average growth rate of firms across all industries.

The need to substantially increase profits thus prompted chemical firms to look at potential areas for expansion, and another input sector, the seed industry, showed the most promise. Large corporate conglomerates, such as the oil company Royal Dutch/Shell, had already moved into this space, but pesticide companies had the advantage of an existing infrastructure for sales to farmers. In addition, genetic engineering technologies to create herbicide tolerant varieties were becoming closer to commercialization, which would allow them to tie proprietary pesticides to seeds—in other words, they would be able to leverage monopolies in one input sector to monopolies in another sector (Kaplow 1985; Harl 2000).

Farmers cannot easily make their own synthetic pesticides, but typically they have the potential to be self-reliant with seeds. The ability of seeds to self-reproduce once posed a formidable barrier to the entrance of large capitalist firms (Berlan and Lewontin 1986; Kloppenburg 2004). This barrier was partially overcome with the development of hybrid corn seeds in the 1930s. Hybrid seeds do not exhibit the same characteristics, such as yield, when they are saved from the previous harvest and replanted, essentially forcing farmers to purchase seeds year after year. Both government and private research investment focused entirely on hybrid corn beginning in 1910, due to this capacity to increase industry profits. Research on open-pollinated corn varieties, which exhibit the parent characteristics when replanted, was abandoned (Berlan and Lewontin 1986). Hybrid sorghum was developed in 1955, and by 1960 approximately 95 percent of corn acreage and 70 percent of sorghum acreage was planted with hybrid seed (Fernandez-Cornejo 2004).

Many other seeds, even those that are heavily subsidized commodities, such as wheat and soy, have been more resistant to hybridization, despite tremendous public and private research efforts (Mascarenhas and Busch 2006). As a result, as late as the 1970s, much of the seed industry in the United States could be characterized as a competitive market, consisting of thousands of firms (Fernandez-Cornejo 2004). Changes in intellectual property rights related to seeds attracted the interest of large corporations, however, and contributed to the demise of many of these mostly small, family-owned businesses. The Plant Variety Protection Act (PVPA) of 1970, for example, granted patent-like protections for sexually reproducing seeds, albeit with exemptions for research and to allow farmers to save seed (Barker, Freese, and Kimbrell 2013). With this strengthening of intellectual property rights, nearly 1,000 formerly independent seed firms were acquired by petrochemical, pharmaceutical, and grain trading firms in the 1970s and 80s (Fowler and Mooney 1990).

By the 1990s, there were just 300 firms in the United States that sold commodity seeds, and the number declined to less than 100 by the end of the following decade (Wilde 2009). At this point, the industry was dominated by agricultural chemical companies, with US CR4 estimates of 80 percent for corn seed, and 70 percent for soybean seed (Schafer 2013). For a brief time pharmaceutical and chemical companies combined to form "life science" companies, but when the expected synergies were not realized, most spun-off their agricultural divisions as separate firms. Globally, from 1996 to 2011 the top three seed firms more than doubled their market share, and achieved oligopolistic control over more than half of the world market for commercial seeds (ETC Group 2013b). The market share figures for leading seed and pesticide firms are presented in Table 7.1, indicating that these top three seed firms are also chemical companies.

The scope of intellectual property protections has been continually expanded via legislation and court decisions in the United States, as shown in Table 7.2. One of the few instances where the United States was not on the leading edge of policy changes that increased the power of dominant firms occurred when the International Union for the Protection of New Varieties of plants (UPOV) was revised in 1991. These regulations prohibited farmer exchange of protected seed varieties, including barter or gift, in contrast to previous agreements (ETC Group 2013b). In 1994, however, the United States amended the PVPA to

Table 7.1 Global seed and pesticide markets, 2011

Seed firm	Market share (percent)
Monsanto (USA)	26
DuPont Pioneer (USA)	18.2
Syngenta (Switzerland)	9.2
Vilmorin/Groupe Limagrain (France)	4.8
	CR4: 58.2
Pesticide firm	**Market share (percent)**
Syngenta (Switzerland)	23.1
Bayer CropScience (Germany)	17.1
BASF (Germany)	12.3
Dow AgriSciences (USA)	9.6
Monsanto (USA)	7.4
DuPont Pioneer (USA)	6.6
	CR4: 62.1

Source: ETC Group (2013b).

Table 7.2 Intellectual property protections for living organisms: key US policy changes

Year	Policy	Impact
1930	Plant Patent Act	Patents on asexually reproducing plants
1970	Plant Variety Protection Act	Patent-like protections for sexually reproducing plants
1980	Bayh-Dole Act	Patents allowed for publicly funded research outputs
1980	*Diamond v. Chakrabarty*	GE organisms patentable
1985	*Ex parte Hibberd*	Plants patentable under general utility patent provisions
1987	*Ex parte Allen*	Multicellular animals patentable
1994	Amended Plant Variety Protection Act	Selling seed without license from owner prohibited
1995	*Asgrow v. Winterboer*	Increased restrictions on selling saved seed
2001	*JEM Agricultural Supply v. Pioneer Hi-Bred*	Saving seeds with utility patents prohibited
2013	*Bowman v. Monsanto*	Enforces contract provisions beyond first sale

match the provisions of the revised UPOV. Many less industrialized countries continue to enforce the less restrictive 1978 version of UPOV, despite pressure from the US government to adopt the most recent version (Aistara 2011; Li et al. 2013).

A 1980 Supreme Court decision, *Diamond v. Chakrabarty*, allowed full patent protections on GE organisms, and soon after, in 1982, the chemical firm Monsanto entered the seed business via its acquisition of the Jacob Hartz Seed Company (Harl 2000). Hartz specialized in soybean seed, and by this time Monsanto was already experimenting with varieties of this crop that could survive direct spraying with their proprietary glyphosate herbicide, Roundup—leaving the plant unharmed but killing the surrounding weeds. The firm anticipated using patents as a way to ensure monopoly profits in the soybean sector via legal enclosure, and to compensate for the lack success with the biological strategy of hybridization. Competing chemical companies also made some seed company acquisitions, but the pace increased dramatically beginning in 1996, when Monsanto commercially introduced Roundup Ready soybeans—the first fully patented, GE crop (Howard 2009c).

Monsanto had benefited from another Supreme Court decision in the previous year, *Asgrow v. Winterboer*, which "signaled a shift in enforcement of plant intellectual property rights from litigation against corporate competitors to lawsuits against the end-user farmer" (Heimes 2010, 109). It also transformed seeds from a fully owned product to a licensed product, making them subject to the provisions of contracts (Aoki and Luvai 2007). With these changes, Monsanto could enlist government support to enforce its contractual restrictions on farmers who bought Roundup Ready seeds, such as the requirement to purchase Monsanto's proprietary herbicide and the prohibition on seed saving (Harl 2000).

The success of the strategy of linking proprietary seeds and herbicides is illustrated by the fact that Monsanto maintained an 80 percent market share for the herbicide glyphosate six years after the patent expired in 2000, despite charging prices that were three to four times higher than generics (Urda 2006). A class-action antitrust suit involving 10,000 farmers, *Pullen Seeds and Soil v. Monsanto*, charged that the firm used anticompetitive tactics to achieve this outcome. These tactics allegedly included rebates to seed companies and dealers to bundle the proprietary herbicide with its seeds, as well as eliminating potential competing technologies through acquisitions. This case, as well as several other suits that made similar claims, were quickly dismissed by the federal court, usually on technicalities such as lack of standing (Dupraz 2012).

Monsanto also aggressively targeted farmers who were believed to have saved their patented seeds. Kem Ralph, for example, was sentenced to eight months in prison and ordered to pay $3 million, although he claimed that he had never signed an agreement, and that the version produced in court was forged (Meek 2006). Monsanto has filed 142 lawsuits involving more than 400 farmers, which had resulted in $23.7 million in recorded judgments by 2012, according to Center for Food Safety (Barker, Freese, and Kimbrell 2013). These figures exclude approximately 4,000 cases brought against farmers that never went to trial, and the resulting settlements were estimated to result in payments to Monsanto of more than $100 million (Barker, Freese, and Kimbrell 2013). The corporation uses anonymous hotlines for neighbors to report suspected violators, as well as private investigators, sometimes even showing the farmer surveillance photos of themselves to underscore their demands (Barlett and Steele 2008).

The practice of saving commodity seeds has dropped dramatically as farmers have adopted GE varieties, attracted by their ability to create culturally desirable weed-free fields with less effort (Levidow and Carr 2007). An estimated 63 percent of farmers saved soybean seeds in 1960, and this level remained high until the mid-1990s, but by 2001, just 10 percent of farmers reported engaging in this practice (Mascarenhas and Busch 2006). The infrastructure for saving seeds is also being lost. Most US counties once had a soybean seed cleaner, for example, but one of the few remaining, Moe Parr, left the business after battling a lawsuit from Monsanto. The judge required Parr to take a sample of every load for testing at Purdue University and then forward the test results to Monsanto to prove that the soybeans were not Roundup Ready (Fraser 2009). The decline in seed saving is resulting in the loss of on-farm breeding efforts that maintain locally adapted varieties and even the knowledge of how to save and replant seeds (Mascarenhas and Busch 2006; Kutka 2011).

With farmers increasingly buying seeds year after year, their prices have risen much faster than other inputs, such as fertilizer and machinery, even though these industries are also highly concentrated (National Farmers Union 2014a). The prices of glyphosate herbicides that are tied to seeds have also risen rapidly (Lin 2013). In addition, the Big Six firms are finding ways to link other chemicals to seed sales. More than 90 percent of the corn seed in the United States and Canada is treated with Bayer's neonicotinoids (which have been implicated in the decline of bee populations) and fungicides, for example, even though studies show no yield gains from this practice (Goulson 2013; Stevens and Jenkins 2014). Another example involves Syngenta's development of a (non-GE) hybrid barley variety—its sales contract forces farmers to also buy a package of chemicals, including fungicides (Blake 2003).

While these tactics have been successful in reducing seed saving and encouraging farmers to buy proprietary chemicals, Big Six firms have increased profits and reduced farmer choices even further by:

- restricting the market availability of previously available varieties, particularly non-patented/conventional varieties and even less expensive GE varieties (Hubbard 2009; Hilbeck et al. 2013)

- presenting an illusion of choice through stealth ownership of regional seed companies (Howard 2009c)

- engaging in a web of cross-licensing agreements with each other, presenting a barrier for other seed firms to access traits desired by farmers (Boyd 2003; Howard 2009c)

These actions are in addition to greater control of the seed industry through numerous acquisitions, which have come at increasingly high prices (Sieker 2009). When Syngenta acquired the biotechnology and seed firm DevGen in

2012, for example, they paid thirteen times the following year's estimated sales. This was a "hefty premium," far above the seed industry norm and was expected to take many years to pay back (Winters 2012). Figure 7.1 shows the hundreds of additional mergers, acquisitions, and alliances that occurred from 1996 to 2013. At the very beginning of this period, Monsanto executive Robert Fraley infamously said, "what you're seeing is not just a consolidation of seed companies, it's a consolidation of the entire food chain" (Horstmeier 1996, 19). He was referring to the critical position of seeds as the first stage in the food system, giving it a structural position of enormous power.

This strategy has paid off for Monsanto, which only entered the seed industry a few decades ago, and is now the top firm, with seeds and seed traits accounting for nearly 80 percent of the growth in its profits (Monsanto 2014). Competing chemical companies have not made as many seed company acquisitions as Monsanto but have similar goals, with Dow stating its intention to increase its revenues from seeds from 5 percent to 50 percent in little more than ten years, for instance (Fatka 2007). Syngenta, which has nearly the same annual revenue as Monsanto but less than one-quarter of which is generated from seeds, has a market capitalization that is 35 percent lower (Nicklaus 2015). Because Syngenta's greater emphasis on chemical sales is not expected to result in the same rate of growth, it has become more vulnerable to takeover by larger competitors.

Seed-chemical firms are also expanding in additional directions, in concert with increasing intellectual property protections that move corporations closer to the hegemony depicted in *Rollerball*. While initial acquisitions focused on subsidized commodity crops, fruits and vegetables are an emerging part of their portfolio, due to policy changes that have allowed more restrictive contracts. Monsanto subsidiary Seminis, for example, has a seal on packages of (non-GE) hybrid Big Beef tomatoes that states, "by opening this container, you agree not to save any seeds, plants, plant parts ... ," and specifies restrictions on breeding, research and seed production (Dillon 2010). Monsanto also acquired germplasm for broccoli that had been bred to be easier to harvest and patented a few hybrid varieties. The company then attempted to extend this patent to *any* variety that had extended crowns. Although the request was denied, they are currently appealing in an attempt to wear down the patent examiner (Hamilton 2014).

DuPont and Monsanto (Box 7.1) were part of the "Intellectual Property Committee" in the 1980s and 1990s, an ad hoc group of thirteen corporations that successfully pushed to extend the US model of intellectual property protections to numerous other countries through multilateral trade agreements

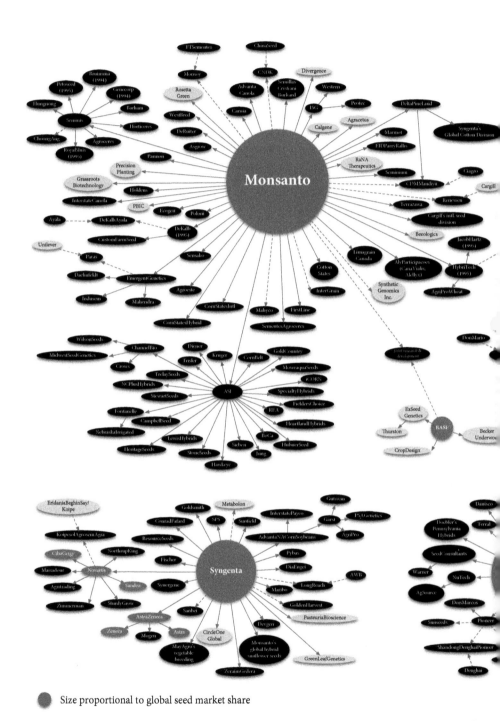

Size proportional to global seed market share

Figure 7.1 Seed industry acquisitions and alliances, 1996–2013.

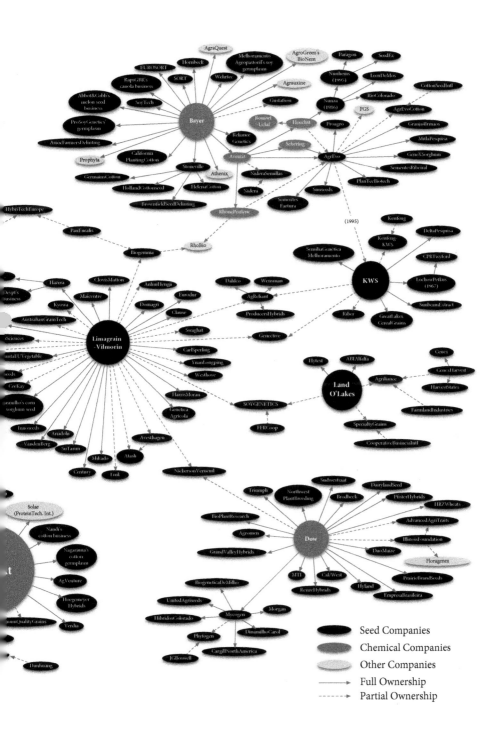

Box 7.1 Co-opetition: DuPont and Monsanto

The interactions between seed giants DuPont and Monsanto are fiercely competitive in some areas, with DuPont going as far as making antitrust claims against its rival. In other areas, however, the two are extremely cooperative, such as working to dismiss critiques of genetic engineering (Lobb 2013) and inflating figures for the amount of land planted with GE crops (Peekhaus 2013). This "co-opetition" is not unusual among dominant firms in the same industry but dramatically illustrates the role the US government plays in mediating these interactions.

The chemical giant DuPont entered the seed industry in 1997 by acquiring a 20 percent stake in the Pioneer Hi-Bred Seed Company, the world's largest seed company at the time. Pioneer was founded by Henry A. Wallace, who sold the first commercial hybrid corn seeds, and went on to become a US vice-president. A full acquisition by DuPont was made two years later, at a price of more than $9 billion. The move may have been a defensive reaction to Monsanto's numerous buyouts in the previous few years. Nevertheless, DuPont has failed to maintain this top position, and they are now second to Monsanto in seed sales and sixth in pesticide sales (ETC Group 2013b).

DuPont filed a private antitrust suit against Monsanto in 2009, charging that their competitor was using a number of illegal tactics to monopolize GE seed traits and stifle innovations (Lim 2013). The DOJ opened an antitrust investigation against Monsanto soon afterwards, but it was very narrowly focused on the minutiae of licensing agreements with competitors rather than a broader exercise of power (Khan 2013a). During this investigation, journalists revealed that DuPont had funded the Organization for Competitive Markets, as well as the American Antitrust Institute, both of which were very vocal in critiquing the dominance of Monsanto and its role in raising seed prices (Neubauer 2009; Corporate Crime Reporter 2010). Interestingly, the DOJ dropped the investigation in 2012 without any public announcement (Philpott 2012). The timing coincided with DuPont, Monsanto, and other seed-chemical companies reaching a voluntary agreement that would purportedly increase farmer access to those GE seeds with expiring patents (Lim 2013). A few months later, DuPont and Monsanto announced that they had also resolved a number of their legal disputes and would cooperate even more closely in sharing patented technologies (Gillam 2013).

Monsanto has frequently been the more aggressive firm in economic, political, and discursive arenas, but DuPont's actions have increasingly followed Monsanto's lead over time. DuPont has recently adopted the tactic of hiring private investigators to spy on farmers suspected of saving seeds (ETC Group 2013a). In addition, DuPont originally had a strategy of alliances with regional seed companies to gain access to their germplasm, rather than the outright acquisitions that Monsanto preferred (Howard 2009c), but by 2010 they began converting many of these alliances into buyouts. A proposed 80 percent stake in Pannar Seeds in 2010 was first contested by the South African government, as it would create a duopoly of two foreign firms in this market (Monsanto already had a dominant share), but DuPont refused to concede and ultimately prevailed in a several year-long legal battle (Pitt 2013).

(Sell 2003). Geographic expansion of the industry via acquisitions and joint ventures has since been rapid, particularly in emerging markets, including China, India, and Brazil. Seed-chemical corporations are also spending billions of dollars to acquire biological pest control and agricultural data analytics firms, raising concerns about how this additional power could be misused, such as by charging farmers more than neighbors for the same product if their yields are higher (Khan 2013c).

With increased government protection of legal exclusions, the biological strategy of hybridization has received less emphasis. More limited enforcement in less industrialized countries, however, has encouraged the development of technologies to prevent seed saving by making seeds sterile, so-called Terminator seeds, or requiring proprietary inputs to remain viable, known as Traitor seeds (Kloppenburg 2010). One of these Terminator traits was jointly developed by the USDA and Delta & Pine Land, a cotton seed company; the firm was quickly acquired by Monsanto after being granted a patent in 1998 (Shand and Mooney 1998). These technologies are not yet commercialized, however, and civil society opposition contributed to a moratorium on their use by 193 countries (ETC Group 2013a).

The chemical industry's takeover of the seed industry has also reduced rates of innovation in this sector and the diversity of the food we eat. Seminis, for example, dropped 2,500 fruit and vegetable varieties, which made up more than one-third of its entire seed catalog, as a cost-saving measure (and to help pay back money borrowed for seed company acquisitions), before being purchased by Monsanto (Dillon 2005). Research and development has shifted to focus on varieties with the potential for blockbuster profits, to the exclusion of those with wider benefits, such as those better adapted to low chemical inputs or specific regions (Schurman and Munro 2010). There is also evidence that dominant firms have expended less effort on innovation when compared to their subsidiaries prior to acquisitions (Schimmelpfennig, Pray, and Brennan 2004). The concentration of germplasm in the hands of a small number of firms has hindered research by others as well. A group of twenty-six anonymous scientists complained about contract restrictions on patented seeds in 2009, which prevented them from working even with commercially available varieties. They stated, "no truly independent research can be legally conducted on many critical questions" (Waltz 2010, 996). Patent-holding firms responded by providing some relief for scientists at agricultural universities, but those at other universities, or who want to examine non-commercialized varieties, continue to face tremendous restrictions (Waltz 2010, 996).

With the loss of numerous varieties and less innovation we are more vulnerable to climate change, diseases, and pests that could drastically impact food crops. One well-known example is the potato blight in Northern Europe, particularly Ireland, which contributed to the deaths of more than 1 million people in the 1840s (Fowler and Mooney 1990). A more recent case was the corn leaf blight that affected the United States in the 1970s and damaged crops worth an estimated $2 billion (Kloppenburg 2004). We have since increased our dependence on a narrow range of crops (Khoury et al. 2014), and the genetic diversity of these crops remains extremely limited. Hybrid corn, for example, accounts more than one-quarter of harvested crop acres in the United States (National Corn Growers Association 2014) and Reed Yellow Dent contributes 47 percent of the gene pool used for creating these varieties (Heinemann et al. 2014).

Imposing homogeneity: livestock genetics

The loss of common resources is also occurring in livestock as a result of increasing powers granted by governments to genetics firms. Monsanto was once involved in the swine genetics industry and, in keeping with its culture of aggressively pursuing monopoly power, filed a dozen extremely broad pig patent claims beginning in 2005. The corporation boldly attempted to extend precedents established by its genetic engineering patents to non-transgenic animals. One patent application raised concerns that any pigs containing a gene supposedly linked to quicker fattening could be claimed as Monsanto inventions. An analysis by Greenpeace indicated that after testing nine different breeds, almost half the pigs they studied contained this genetic variation (Then 2005). The European Patent Office, despite facing public demonstrations, granted the patent, albeit with a narrower scope. By that point, however, Monsanto had sold its swine breeding division (Peter 2009). The episode signified a shift in the mostly privately held livestock breeding industry, away from trade secrets and biological strategies of enclosure and toward government enforcement of intellectual property protections (Then and Tippe 2011).

Animal genetics is one of the smallest input sectors, with global annual sales of approximately $4 billion (Fuglie et al. 2011). For comparison, the seed industry has estimated yearly sales of $35 billion and pesticides at $44 billion (ETC Group 2013b). Control of animal genetics for a number of livestock species is extremely concentrated, however, which has contributed to an enormous narrowing of the genetic diversity of livestock. The technology treadmill and the demands of increasingly industrialized processors and retailers have helped persuade farmers to accept more homogeneous animals. Large commercial breeders often ignore susceptibility to disease to focus on traits that increase profits and instead rely on antibiotics to control illness. This practice contributes to antibiotic resistant microbes and threatens human health. It also increases the risk that an epidemic disease will have a dramatic impact on the food supply, perhaps even greater than recent outbreaks of avian influenza in poultry (Alders et al. 2013), or new coronaviruses in swine (Wang, Byrum, and Zhang 2014).

With the development of artificial insemination and hybridization in the 1940s, farmers and ranchers began relying on off-farm breeders, raising barriers to entry for small breeding firms. Hybridization was first applied to poultry by Pioneer's Henry Wallace, with methods he had developed in the commercial corn sector. As with seeds, the offspring do not exhibit the same desired traits, such as "high number of eggs per year, high feed conversion, fast growth, (or) high percentage of lean meat" (Gura 2007, 15). Embryo transfer technologies were then developed in the 1970s, followed by genetic marker technologies the 1990s, and sexed semen in the 2000s (Hasler 2003). Cloning is not heavily commercialized yet, as it is expensive, and the resulting animals are often born with numerous disorders (Chug 2011). The US FDA, however, approved the sale of unlabeled meat and milk from cow, pig, and goat clones and their offspring in 2008.

In the case of poultry breeding markets, just two or three firms control 94 percent or more of sales, as many smaller firms have been acquired, or exited these industries in recent years (Table 7.3). In addition, two firms, Erich Wesjohann (EW) Group and Hendrix Genetics, are dominant across multiple poultry sectors. Both firms are privately held, with EW Group based in Germany and Hendrix Genetics in the Netherlands.

Table 7.3 Global livestock genetic markets

Turkeys	Laying hens	Broilers	Swine
EW Group	EW Group	EW Group	Genus
Hendrix Genetics	Hendrix Genetics	Tyson	Hendrix Genetics
		Groupe Grimaud	Groupe Grimaud
			Smithfield
CR2: 99%	**CR2**: 94%	**CR3**: 95%	Four firms control 2/3 of research & development

Source: ETC Group (2013b).

Estimates of effective population sizes for different breeds of chickens are difficult to uncover, because they are considered trade secrets, but the FAO suggests the majority of commercial strains are based on just four breeds (Gura 2007). In the United States, most broilers are Cornish Rock crosses, and most eggs are from White Leghorns. For turkeys, there is only one breed, the Broad Breasted White, that makes up nearly 100 percent of the global supply, and acquisitions in this sector have reduced this breed to just a small number of strains. All of the breeding stock in the United States was destroyed in 2004, after the acquisition of Nicholas by EW Group's subsidiary Aviagen, for example (Walker 2009). The Broad Breasted White was selected for its high proportion of white meat that is preferred by consumers, but the birds cannot reproduce without artificial insemination and have difficulty walking due to the disproportionate weight of the breast (Martrenchar 1999). The American Livestock Breed Conservancy conducted a census in 1997 and found only 1,335 individual birds in the entire United States that did not belong to this breed (Walker 2009).

Poultry is followed closely in its level of concentration by swine genetics. This sector has also experienced a number of significant mergers and acquisitions in the last twenty years, and there is now greatly increased reliance on artificial insemination. Due to vertical integration, one of the largest swine genetics firms is a subsidiary of Smithfield (now owned by China's Shuanghui), which only supplies that firm's operations (Johnson 2006). Although there are more than 500 breeds of swine, the majority in the United States are hybrids derived from just three (Duroc, Hampshire, and Yorkshire). The effective population size for each of these breeds is less than 100, which is considered the minimum for maintaining genetic diversity (Blackburn, Welsh, and Stewart 2005).

Dairy cattle breeding is also quite concentrated. There are approximately 1,000 breeds of cattle worldwide, but just one breed, the Holstein from the Netherlands, makes up more than 85 percent of milking cows in US (Covington 2013). Holsteins are popular due to high milk production and the ability to tolerate high grain diets but have lower butterfat and are more fragile than many other breeds (Kneen 1993; Kurlansky 2014). There are approximately 9 million Holsteins in the United States, but over 60 percent of them are from just four family lines, with an effective population size of just 39 animals (Hendrickson and Heffernan 2003; Blackburn, Welsh, and Stewart 2005). Scandinavian countries have recognized the risks of such uniformity in dairy cows and have established breeding programs that accept less productivity in order to increase genetic diversity (Gura 2007).

Beef cattle have to this point been less suited for capitalist involvement because production is less confined (and therefore reproduction is less controlled), and they have longer gestation periods than poultry or swine. Currently, this results in a lower potential for intellectual property protection, although future technological developments could attract more interest from dominant firms. For cattle breeding there are monopolies in individual coun-tries, but no leading firm has a market share of more than 25 percent in the EU (Fuglie et al. 2011).

Many of the leading animal genetics firms have expanded into the aquacul-ture breeding business in recent years as a result of government investments in research, which focuses on shrimp, tilapia, salmon, and rainbow trout (Asche 2008). EW Group acquired a majority stake in Norway's AquaGen; Hendrix acquired two Scottish salmon breeders; and Groupe Grimaud (Box 7.2) has a shrimp subsidiary called Blue Genetics (ETC Group 2013b). The US firm

Box 7.2 Too Global to Fail? Groupe Grimaud

Groupe Grimaud of France is the second largest multi-species animal genetic company in the world, after Hendrix Genetics. It is 70 percent owned by Frédéric Grimaud and family, 30 percent by financial institutions (Gura 2007). Grimaud started as a duck breeding company, but due to slow growth in this industry the firm moved to acquire the much larger Hubbard in 2005. At the time Hubbard was the world's third largest broiler firm, having acquired Shaver and ISA but was losing money. Grimaud's restructuring focused on reducing the number of genetic lines and centralizing global operations at one loca-tion in France, but this backfired when the avian influenza epidemic of 2006 resulted in a loss of half the business (van der Sluis 2012). The firm now has operations in numerous other countries, including Italy, Poland, Netherlands, US, China, and Malaysia (Grimaud 2014).

To diversify risks and expand into new markets, Grimaud acquired the swine breeding firm, Newsham Choice Genetics, in 2010. Newsham made a number of acquisitions before being acquired itself, including Monsanto Choice Genetics, DeKalb Swine, Seghers, and Ausgene (Fuglie et al. 2011). Grimaud actually facilitated the Newsham acquisition of Monsanto Choice Genetics in 2007, before eventually taking over the firm (van der Sluis 2012). Since then, Grimaud has acquired a majority stake in the swine genetics firm, Pen Ar Lan—also headquartered in France with additional operations in Poland, Brazil, and Canada—and rebranded the swine business as Choice Genetics in the United States. Although its rapid growth was aided by increasing intellectual property protections on animal breeding, enforcement of these laws also contributed to the bankruptcy of its Choice Genetics subsidiary in 2014, after it lost a dispute with the animal health firm Scidera over unpaid royalty payments (Chutchian 2014).

AquaBounty has developed hybrid fish, as well as a patented, GE salmon, for which it has been seeking commercial approval for nearly two decades. This Atlantic salmon contains genes from both Chinook salmon and ocean pout and grows faster than conventional farmed salmon. Although major retailers, such as Safeway and Kroger, have stated that they will refuse to carry it, the company has received numerous government subsidies, such as half a million dollars from the USDA to research sterility technologies (Leschin-Hoar 2011; Goeden 2014).

The US and EU governments have also provided generous subsidies across numerous other livestock breeding sectors, despite negative impacts on diversity, animal welfare, human health and the environment (Gura 2007). The United Nations Food and Agriculture Organization estimates that 20 percent of livestock breeds are at risk of extinction due to the emphasis on large-scale, concentrated systems (Pilling and Rischkowsky 2007). The animal genetics industry's emphasis on highly profitable traits and homogeneity has challenged their capacity to respond to changing conditions in industrial livestock production, such as retailer demands for larger cages. It also creates difficulties for farmers and ranchers seeking livestock adapted to pasture production, because dominant firms primarily offer those that have been bred for climate-controlled and confined systems (Gura 2007).

Reclaiming the commons?

As with other stages of the food system, antitrust has not been an effective route for slowing concentration in agricultural input industries in recent decades. National governments are heavily involved in facilitating increased enclosure of plant and animal genetics commons and excluding more of the world's population from access to these resources. Civil society has attempted to resist by demanding an end to life patents, in order to reclaim the common heritage that is being privatized through this strategy (Then and Tippe 2009). Although patents on human genes were struck down by the US Supreme Court in 2013, the precedent that was set in the 1980s for other organisms has yet to be overturned. There have been many skirmishes over the scope of patent claims, however, with some being denied or narrowed, as with Monsanto's pig patent. Not surprisingly, there have also been many failed attempts to challenge the expanding power of dominant firms.

The case of Vernon Bowman, which went all the way to the US Supreme Court, was one such failure. Bowman was a soybean farmer from Indiana who engaged in the long-standing practice of buying soybeans from a grain elevator and planting them. With the rapid adoption of Roundup Ready seed, many of these were glyphosate tolerant by 1999. That year he applied the herbicide to his fields and saved and replanted the seeds over eight growing seasons. He did not believe Monsanto's intellectual property claims applied to his actions, due to the legal "first sale doctrine" or patent exhaustion, which limits rights of patent holders to the original purchase. Bowman, therefore informed Monsanto of his actions, and was sued in 2007. Monsanto's large team of lawyers, which included a former US solicitor general, prevailed over the bankrupt Bowman at every stage in the legal process and ensured that patent exhaustion did not apply to the firm's seeds (Lim 2013).

Another effort to reclaim the commons is led by a group of University of Wisconsin professors to create "open source" seeds. They are attempting to use intellectual property laws to secure greater access to these resources, reversing the typically restrictive nature of these statutes (Sussman 2014). The development of open source intellectual property rights has the potential to impede the patenting of plant genetic material and the increasing appropriation of germplasm, by using the tool of government enforcement to preclude monopolization. This will require tremendous cooperation from farmers, researchers, and activists. In addition, it can be expected to face strong opposition from both governments and the seed oligopoly, but it may provide a small space of resistance (Kloppenburg 2010, 2014). The first examples of open source seeds, two varieties of carrots, were released in 2014 (Sussman 2014).

Intellectual property protections do not yet apply to open-pollinated (or heirloom) seeds and heritage breeds of livestock. Private firms and nonprofit organizations specializing in these areas are small but experiencing extremely rapid growth. The 2008 economic crisis revitalized interest in gardening, and along with it, the advantages of seeds that are not hybrids. An example is the Baker Creek Heirloom Seed Company, with operations in three different states and more than 100 employees. The company prints and distributes several hundred thousand free copies of its full color catalog each year, which raises awareness of varieties that are close to being lost. It also sponsors an Heirloom Expo in Santa Rosa, California each year that attracts 20,000 people. Seed Savers Exchange in Iowa is another effort to preserve not only biological diversity but also the sociocultural heritage of heirloom varieties (Carolan 2007). Dissatisfaction with this organization's increasing bureaucracy, however, has engendered a new Grassroots Seed Network (Pols 2014).

Numerous public libraries in the United States have started "seed libraries" and ask if you "check out" a packet of seeds to return harvested seeds for other patrons to use following the growing season (Hartnett 2014). Beginning in 2014, however, several US states, including Pennsylvania and Minnesota, have shut down seed libraries for their failure to conduct expensive germination tests. Although some legal analysts suggest these actions are a misapplication of current regulations, several states have proposed legislation to exempt seed sharing from such unnecessary restrictions (Cook 2015). Such efforts are also more difficult in the EU, which has a very restrictive seed registry. It prohibits varieties that have not gone through the expensive certification process from being sold, or even exchanged with neighbors (Kastler 2005), although one company, Real Seeds, has thrived by focusing on an "amateur" exemption in the UK for the home garden market.

Public interest in heritage breeds of livestock is also seen as a threat by powerful organizations, which are manipulating fears of invasive species or epidemic diseases as justification to discourage alternatives to the industrial model. Niche markets are spurring the demand for livestock breeds that are more adapted to foraging, or have different taste characteristics than industrial breeds. There has been a resurgence of interest in heritage breed turkeys, for example, and the US population is estimated to have increased to more than 10,000 by 2006 (Walker 2009). In Michigan, the Department of Natural Resources developed regulations that essentially outlawed heritage swine on the basis of physical characteristics, such as straight tails, with no genetic criteria. Resorting to the weak justification that it was necessary to control feral pigs, heritage swine farmers were ordered to kill their fenced livestock. This move was strongly supported by the Michigan Pork Producers Association, which represents large-scale operators. Mark Baker was one of the few farmers to resist the order, and after a long and expensive battle achieved an exemption, but only for his farm (Gumpert 2014).

In a similar fashion, the risk of salmonella is touted as a reason to restrict outdoor poultry production, including organic production, despite a lower incidence of the disease in less confined flocks (Cornucopia Institute 2013). Another example was detailed by Linda Faillace in her book *Mad Sheep* (2006). She experienced government surveillance and was forced to give up her East Friesian and Beltex sheep due to USDA claims that they could harbor scrapie, related to bovine spongiform encephalopathy (BSE), or mad cow disease. After the 125 animals were slaughtered, however, they were found to be disease free.

The often heavy-handed responses to efforts to reclaim the commons have led to more public discussion of the motives of government agencies, as they reveal the interests that are truly being protected by their actions. Even people who are not directly involved in food production are expressing concerns about losing access to seed and breed resources (van Bommel and Spicer 2011). In addition, broader concerns about safety have helped form large coalitions to oppose enabling technologies, such as genetic engineering and cloning, and slowed their commercialization (Schurman and Munro 2010). It is also important to remember that a significant percentage of the global population has remained largely outside the reach of the new enclosures. Approximately 3 billion people engage in indigenous or peasant production, which does not show up in market share figures, including as much as 90 percent of the seed planted in the Global South (ETC Group 2013b). This represents territory into which dominant firms could continue to expand, however, unless enclosures (including land grabs) are slowed much more substantially.

This chapter characterized recent changes in input industries, the critical first stage of the food system. Common resources, including agricultural plants and animals, are increasingly privatized, with negative impacts for farmers and the future of food supplies. This process could not occur without governments redefining and enforcing property rights that strongly favor dominant firms. Government efforts were also essential for increasing capitalist involvement in the organic food industry, as described in the next chapter. A key difference in the approach was that it involved control over a voluntary set of standards.

Chapter 8

Standardizing resistance: the organic food chain

> *"Listen, this whole system of yours could be on fire and I*
> *couldn't even turn on the kitchen tap without filling out a*
> *twenty-seven B stroke six ... bloody paperwork."*
> —Archibald "Harry" Tuttle (*Brazil*)

When public initiatives to label genetically engineered (GE) foods were placed on the ballots in California in 2012, Washington State in 2013, and Oregon in 2014, early polls showed overwhelming public support. All three suffered narrow defeats, however, after transnational corporations and their trade associations poured millions of dollars into anti-labeling advertising campaigns in the weeks before the votes (Giroux 2013). Interestingly enough, some of these corporations had organic food divisions. Organic standards prohibit the use of GE organisms, and many consumers seek out organic food specifically to avoid them (Equation Research 2005). Because the organic food brands of companies like Coca-Cola, Kellogg, and General Mills were acquired by stealth ownership, with no indication of their corporate parentage on the packaging, organic consumers may have unknowingly contributed to funding these defeats.

This is just one example of dominant food firms co-opting a movement that resisted their values and transforming it into another means of increasing their power (Jaffee and Howard 2010). How were they able to do so for the organic movement? One key strategy was to gain an increasing amount of influence over the standards that define the very meaning of "organic." Lawrence Busch calls standards "recipes for reality" (2011a, 73). They delineate what is acceptable or unacceptable in many aspects of society but often go unnoticed or fade into the background for most of us (Busch 2011b). A closer look at standards can help illuminate social relationships that are often invisible by examining who has the most control over the process of setting them and who benefits the most from the results.

This chapter analyzes the impacts of the evolution of organic standards in more detail throughout key segments of the food system, including retail, distribution, processing, production, and agricultural inputs. Although rapid growth affected each of these segments in different ways, some "buying in" has occurred in all of them to an extent, as capitalists have adapted to the requirements of organic standards and reduced some of the ecological impacts of conventional agriculture. Some "selling out" has also taken place, however, as a number of original ideals of organic practice have been removed from the meaning of the term, despite strong resistance from activists. Organic is becoming less of an alternative to the conventional food system and is increasingly dominated by larger and more profit-focused firms, particularly in the off-farm segments. This concentration is well hidden, as the values of organic consumers encourage gigantic firms to conceal their ownership after making buyouts. These changes have motivated efforts to create a number of newer alternatives to organic, with an emphasis on criteria omitted from the USDA organic standards.

Harmonizing meanings of "organic"

Organic standards now differ from those that were in effect through the late 1990s, when conventional food firms had far less involvement in the industry. One of the barriers for very large firms was a "patchwork" of at least fifty-five state and regional organic standards, which had numerous similarities but also some important differences (Fetter and Caswell 2002). These differences posed a challenge for firms seeking to sell processed organic products with ingredients sourced from more distant locations, as some certifiers did not recognize others' standards if they lacked equivalency agreements (Keupper 2010). After the state of California became involved in overseeing organic standards in 1990, pressure to bring this model to the national level quickly followed (Guthman 1998).

The national legislation was passed in 1990, but the USDA, which was charged with implementing the standards, did not release a draft until 1997. This draft was highly controversial, as it redefined "organic" to allow genetic engineering, irradiation, and sewage sludge—the so-called "Big Three"—although these had always been excluded from organic food production. The proposal was a clear attempt to change the widely held meaning of organic in order to facilitate the entrance of large-scale capitalists into this rapidly growing and profitable sector (Allen and Kovach 2000). The USDA received a record number of public comments on the draft (more than a quarter million people), the vast majority opposed to the inclusion of the Big Three (Vos 2000). Due to this resist-ance, the legitimacy of the draft was challenged, and the USDA was pressured to release a final version that was closer to the California standards—prohibiting the Big Three, along with synthetic pesticides and fertilizers, growth hormones, and antibiotics. This national standard, which lacked some of the deeper, origi-nal meanings of organic, such as local marketing of less processed foods, was then phased in from 2001 to 2002.

This change significantly contributed to consolidation in organic food indus-tries and likely contributed to more rapid growth as well (Obach 2015). By 2013, it was estimated that organic food sales in the United States were $32.3 billion, which amounted to more than 4 percent of the total food market (Haumann 2014), and global sales exceeded $60 billion (Soil Association 2013). Although growth has slowed since the 1990s, annual sales are still rising faster than the rest of the food industry, suggesting that the market share for organic foods will continue to increase in the near future.

Conflating natural and organic: retailing and distribution

When interest in alternatives to the conventional food system increased in the 1970s and 1980s, there were not enough supplies of organic foods to stock an entire store. Retailers and distributors coalesced around the much broader category of "natural" foods by necessity. Whole Foods, Inc., for example, could only stock 2–5 percent organic produce in 1980 but was able to increase this to 50 percent by 1989 (Dobrow 2014). Although less formal than the definition of organic, a standard also developed around the meaning of natural food. This was largely shaped by retailers, particularly chain stores, although consumer feedback also played a part. Whole Foods, for instance, banned artificial colors, flavors, and preservatives when it was founded (Kowitt 2014). Its list of prohibited substances had expanded to seventy-eight by 2014 (Whole Foods 2014). Josée Johnston (2008) notes that Whole Foods continues to use the words "natural" and "organic" interchangeably in their marketing efforts, making it difficult for customers to distinguish the two. One result of this conflation is that the retailing and distribution segments of the organic food chain presented fewer barriers to large-scale firms and consolidated even earlier than other segments.

Natural foods retailing was heavily influenced by members of a food coun-terculture, due to their dissatisfaction with choices at conventional retailers (Belasco 2007). Many of them formed their own cooperative buying clubs and stores to obtain desired products; more than 5,000 food co-ops were founded in the 1970s, adhering to principles including democratic control and the return of surplus to members (McGrath 2004). By the 1980s, these innovations had been copied by more capitalist-minded entrepreneurs, who frequently met to socialize and exchange business information. In the 1990s, members of this group opened multi-store chains and bought out their smaller competitors (Dobrow 2014). They gained an advantage over co-ops and smaller private retailers through their access to capital and the ability to offer more product choices to customers.

Eventually two of these firms, Whole Foods and Wild Oats became large enough to be publicly traded and, with funding from even larger capitalists, were able to dominate the new niche of natural and organic foods. Along with the chain, Trader Joe's, they accounted for nearly one-quarter of organic food sales by 2006 (Supermarket News 2007). Whole Foods and Wild Oats had each acquired more than a dozen competitors before themselves combining: Whole Foods, with 190 stores, took over the majority of Wild Oats' 112 stores in 2007. The number of food co-ops had declined drastically by this time to approximately 300, mostly very small, stores and accounted for a much lower proportion of sales.

A similar transition occurred in the natural/organic distribution system. Cooperatively owned distributors, which once supplied retail food co-ops at a regional level, were driven out of business by private distributors, including two that eventually gained a national scope (see Box 8.1). By 2002, United Natural Foods Incorporated (UNFI) and Tree of Life were estimated to control

---

Box 8.1 Conquering the Cooperatives: United Natural Foods Incorporated

The rise of UNFI coincided with the demise of more than two dozen cooperatively owned natural food distributors in recent decades, many of which were acquired by this publicly traded corporation. Although UNFI was originally a privately held distributor, executives recognized the benefits to be gained by accessing money from Wall Street investors. This funding enabled the firm to keep pace with the rapidly growing natural foods industry by adding to its infrastructure, such as warehouses, refrigeration, and trucks.

Many of UNFI's competitors, particularly those organized as cooperatives, were unable, or unwilling, to expand at the same pace. As a result, the number of cooperative distributors in the United States declined from twenty-eight in 1982 to three by 2002 (Gutknecht 2003). In that year, when the national organic standards went into effect, UNFI acquired the largest two remaining cooperative distributors, Blooming Prairie and Northeast. In addition, North Farm co-op went bankrupt, despite previously merging with Michigan Federation and Common Health co-ops. By 2008, the last remaining cooperative distributor in the United States was Frontier, which had nearly failed in the early 2000s (Howard 2009a). One key to Frontier's continued existence is its focus on spices and teas, which are both lightweight and expensive compared to other foods. This allows them to cost-effectively ship products through Federal Express, rather than create their own transportation infrastructure.

The relationship between UNFI and retail co-ops is sometimes contentious because the distributor gives Whole Foods—by far their largest customer— better prices and delivery schedules. When UNFI's unionized drivers in Washington State went on strike in December 2012, they were supported by several area retail co-ops. The Olympia Food Co-op went further and withheld orders from UNFI, but the boycott only lasted one week since, like many co-ops, it was dependent upon the distributor for more than 50 percent of its packaged products (Olympia Food Co-op 2013). UNFI immediately replaced almost half of the striking employees, but two months later agreed to a wage increase and the reinstatement of these workers (Paul 2013). The deal did not appear to hurt the corporation, which increased its market capitalization to nearly 4 billion dollars by the end of 2014 nor its executives; the chairman of the board, Michael Funk, founded Mountain People's Warehouse out of his Volkswagen van in 1976 but now lives in a palatial off the grid home, on 1,200 acres in California (Lawrence 2011).

approximately 80 percent of the distribution of processed organic foods (Sligh and Christman 2003). Tree of Life failed to grow as quickly as its main competitor, however, and was acquired by a smaller distributor, Kehe, in 2010. Although both Whole Foods and UNFI are experiencing increased competition as a result of the mainstreaming of organic food—including from Walmart and Sysco—their overwhelming dominance in retailing and distribution is a bottleneck for many organic food processors. One analyst noted that some manufacturers must sell a significant percentage of their products through this pipeline and that "they can make or break you" (Boyle 2008).

Whole Foods and UNFI now have a large say in what natural means, taking over a role that was once held by Mrs. Gooch's. The southern California retail chain, named after founder Sandy Gooch, had some of the most rigorous standards in the natural foods industry. It was also one of the biggest firms in this segment in the 1980s and early 1990s, before being acquired by Whole Foods in 1994. A common question retailers and distributors asked of manufacturers was "Is it Goochable?," with some of them asked to reformulate their products in order to meet the de facto industry standard (Dobrow 2014, 122). This pre-screening of products by retailers allowed consumers to trust that objectionable ingredients had already been excluded, a process the industry calls "choice editing." Although needing to be prodded by customers at times, Whole Foods has been on the leading edge of retailer choice editing, particularly with respect to GE ingredients in the United States. The retailer removed these ingredients from their private label products in 2009 and promised to label all products containing them by 2018 (Robb and Gallo 2013).

Whole Foods and UNFI have power that extends beyond what ingredients are included. Whole Foods' buyers, for example, may demand that suppliers change their names, develop new products or make design modifications as a condition of being carried on their shelves (Gasparro and Josephs 2015). These two firms can also decide whether or not natural/organic innovations will be pushed to much broader markets or languish in obscurity, as with their role in the growth of a fermented tea beverage, kombucha (Boyle 2008). Increasing competition from conventional firms, like Walmart and Kroger, will not significantly reduce this power, because they tend to follow the lead of the "ultimate gatekeeper," Whole Foods (Gasparro and Josephs 2015). Retailing, however, is experiencing the re-entrance of competition from smaller natural foods chains. These include several started by entrepreneurs who had previously sold out to Whole Foods/Wild Oats, including Sprouts Farmers Market in Phoenix, Arizona, New Seasons in Portland, Oregon, and Alfalfa's in Boulder, Colorado.

Stealth ownership, stealth shaping of standards: packaged foods

Investors typically like to be informed when a company adds to its portfolio of brands in a rapidly growing segment of the food industry. General Mills, for example, made public announcements when it acquired the organic brands, Cascadian Farm and Muir Glen, in 1999, Larabar in 2008, and Food Should Taste Good in 2012. In 2013, however, when the firm acquired Immaculate Baking, a producer of natural and organic refrigerated doughs and pie crusts, it was only briefly noted later that year in General Mills' annual report. This illustrates a long pattern of increasing corporate ownership of organic packaged foods, and increasing concealment of these trends.

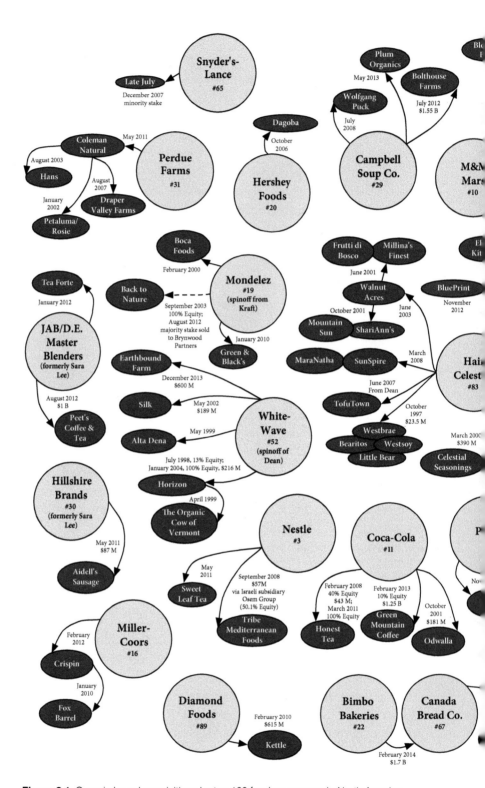

Figure 8.1 Organic brand acquisitions by top 100 food processors in North America.

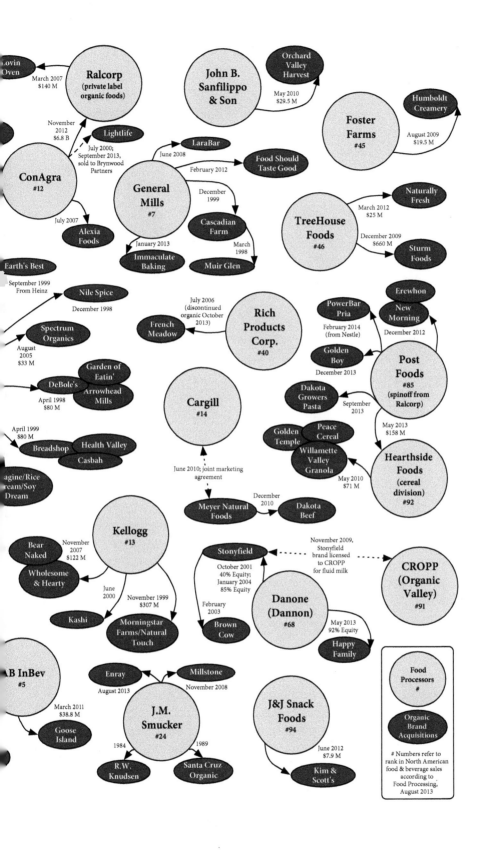

Lovin' Oven

March 2007
$140 M

Ralcorp
(private label organic foods)

November 2012
$6.8 B

Lightlife

July 2000;
September 2013,
sold to Brynwood
Partners

ConAgra
#12

July 2007

Alexia Foods

John B. Sanfilippo & Son

Orchard Valley Harvest

May 2010
$29.5 M

LaraBar

June 2008

Food Should Taste Good

February 2012

December 1999

General Mills
#7

Cascadian Farm

January 2013

Immaculate Baking

March 1998

Muir Glen

Foster Farms
#45

August 2009
$19.5 M

Humboldt Creamery

TreeHouse Foods
#46

March 2012
$25 M

Naturally Fresh

December 2009
$660 M

Sturm Foods

Earth's Best

September 1999
From Heinz

Nile Spice

December 1998

Spectrum Organics

August 2005
$33 M

DeBole's

April 1998
$80 M

Garden of Eatin'

Arrowhead Mills

April 1999
$80 M

Breadshop

Health Valley

Casbah

Imagine/Rice Dream/Soy Dream

Rich Products Corp.
#40

July 2006
(discontinued organic October 2013)

French Meadow

Cargill
#14

June 2010; joint marketing agreement

Meyer Natural Foods

December 2010

Dakota Beef

PowerBar Pria

February 2014
(from Nestle)

Erewhon

New Morning

December 2012

Golden Boy

December 2013

Post Foods
#85
(spinoff from Ralcorp)

Dakota Growers Pasta

September 2013

May 2013
$158 M

Golden Temple

Peace Cereal

Willamette Valley Granola

May 2010
$71 M

Hearthside Foods
(cereal division)
#92

Kellogg
#13

Bear Naked

November 2007
$122 M

Wholesome & Hearty

June 2000

Kashi

November 1999
$307 M

Morningstar Farms/Natural Touch

Stonyfield

November 2009,
Stonyfield brand licensed to CROPP for fluid milk

October 2001
40% Equity;
January 2004
85% Equity

February 2003

Brown Cow

Danone (Dannon)
#68

May 2013
92% Equity

Happy Family

CROPP (Organic Valley)
#91

AB InBev
#5

March 2011
$38.8 M

Goose Island

Enray

August 2013

Millstone

November 2008

J.M. Smucker
#24

1984

R.W. Knudsen

1989

Santa Cruz Organic

J&J Snack Foods
#94

June 2012
$7.9 M

Kim & Scott's

Food Processors
#

Organic Brand Acquisitions

Numbers refer to rank in North American food & beverage sales according to Food Processing, August 2013

Nearly one-third of the 100 largest food processors in North America have acquired pioneering organic brands (Figure 8.1), but very few disclose these ownership ties on packaging or brand websites. General Mills, for example, does not display their big "G" symbol on any of the products noted above. Many of these brands continue to highlight stories of their founding, to give the impression that they remain small and idealistic. Cascadian Farm's logo, for example, shows an image of the original farm along with the text "since 1972" and "founded in Skagit Valley, WA." A closer look at the ingredients, however, indicates that many of them are not sourced from this location, but as far away as China—including their "California" mix. Kashi was acquired by the giant food processor Kellogg in 2000, but until 2013, the "meet us" page on its website stated: "We are a small (after twenty-five years, still fewer than seventy of us) band of passionate people ... " Mary Muiry, an executive at the retailer Wild Oats, explained the reasons for concealing buyouts by conventional food companies, stating: "Our most loyal customer is somewhat leery of established brands" (Helliker 2002).

This leeriness is probably justified, given that large firms have influenced the national organic standards to their benefit, and many organic consumers would find these changes objectionable if they were aware of them. Some key modifications have been the result of pressure from the Organic Trade Association (OTA), which receives the majority of its funding from the industry's largest firms. A particularly notable example was the battle over synthetics in processed foods. Arthur Harvey, an organic farmer who sued with his standing as an organic consumer, argued that the intent of the national organic legislation was not being followed in a number of areas. In 2005, he won on appeal with his several of his arguments, including his opposition to the use of synthetic additives, such as cellulose for anti-caking. Many large organic processors relied on synthetics by this time, and they mobilized quickly to respond to this decision. Later that year, the OTA convinced Congress to anonymously insert a last minute rider to the Agricultural Appropriations Act that altered the standards—a stealth move that essentially nullified Harvey's court victory (DuPuis and Gillon 2009).

There is constant pressure behind the scenes to change standards in ways that favor powerful interests but also the ever present risk that if pushed too far or too visibly, consumer confidence in the meaning of organic could be weakened. The allowance of up to 5 percent non-organic ingredients in products that feature the USDA organic seal is one area of contestation, with recent controversies over the use of conventional hops in beer (removed from the approved list in 2010), intestines for sausage casings, and carageenan (a seaweed-derived thickener/stabilizer) in numerous products. Another site of conflict involves the National Organic Standards Board (NOSB), which is designated to provide volunteer citizen oversight of changes to USDA's organic standards. The composition of this fifteen-member board has shifted since its inception, with more representation from very large firms. In one case, an executive of General Mills, Katrina Heinze, was appointed to fill one of three designated "consumer/ public interest advocate" seats (Cummins 2006). As a result of public criticism, led by Consumers Union and the Organic Consumers Association, Heinze declined the appointment, and instead this seat was filled by a former General Mills employee. The next year, however, Heinze accepted the lone "scientist" seat, with representatives of big firms appointed to several others, including one designated for "environmentalists/resource conservationists."

There have been other negative impacts arising from the stealth acquisitions of organic brands by much bigger food companies. Common outcomes include a weaker commitment to organic ideals, or even dropping organic food production from brands that have historically been associated with organic ideals. Examples of products that now contain little to no organic ingredients include those of Coca-Cola's subsidiary Odwalla and WhiteWave's subsidiary Silk. Arran Stephens of Nature's Path, who has been in the industry since the 1960s, is not surprised by changes like these; he once sold a natural food company called LifeStream, and later bought it back from Kraft. Stephens (2008) said that he has consistently seen the "soul gutted out" of organic food companies after buyouts.

Not all organic food firms have sold out to bigger firms or venture capitalists (Box 8.2), which is remarkable considering that many receive frequent offers, at prices that are much higher than average for the food industry. Gary Erickson, CEO of Clif Bar, for example, nearly sold his company to Quaker Oats/Pepsi. At the last minute he decided to walk away from their $120 million offer, as detailed in his book *Raising the Bar* (2004). There are more than a dozen pioneering organic firms with tens of millions dollars in annual sales that have also remained independent (Table 8.1). Refusing such offers is particularly noteworthy in view of the fact that these firms are now competing against some of the largest food companies in the world. In Erickson's case, he was drawn to the negotiating table after seeing two of his largest competitors in the energy bar industry acquired by Nestlé and Kraft. Size results in tremendous power, such as the ability to cross-subsidize, or shift revenues generated from other parts of the firm in order to drive independent competitors out of business. These strategies include temporarily dropping prices below the cost of production, or spending huge amounts of money on marketing and advertising. Larger firms also have

Box 8.2 Bundling Brands: Venture Capitalists

A consistent pattern in the concentration of the organic food industry has been the involvement of investment firms, more commonly known as venture capitalists (Howard 2009b). These investors make acquisitions with a goal of beating the average rate of return by wide margins, typically 25–100 percent compounded annually, to be paid out within three to seven years (MSI 2007). Achieving this usually requires either selling a firm to a much larger corporation, or transforming it into a publicly traded company. An example of the latter was Solera Capital, which acquired several pioneering organic brands, including Annie's Naturals salad dressings and the separate Annie's Homegrown pastas, beginning in 2002. After increasing distribution to retailers like Target and Costco, an initial public offering ten years later, under the Annie's, Inc. name, resulted in a payout to Solera Capital nearly seven times higher than their investments (Dezember 2012). By 2014, however, shareholders received another big windfall when General Mills acquired Annie's Inc. for $820 million. This was the second highest valuation relative to annual earnings ever recorded for a food company, just behind Ben & Jerry's takeover by Unilever in 2001 (Bamber 2014).

A key strategy in buyouts is "bundling" together many previously competing products within the same category. The Charterhouse Group, for example, described the organic industry as fragmented, and saw the potential of "developing and implementing master consolidation plans" in the organic bread sector (Charterhouse Group 2005). After buying three organic bread firms in 2005, they sold them nine years later for $61.3 million to the organic foods juggernaut Hain Celestial (Watson 2014). Hain Celestial had itself received investments from several venture capitalists, and has acquired dozens of brands to consolidate numerous organic food segments. Booth Creek Management Corporation, which includes as a partner the CEO of Swift & Co. (George Gillette), acquired a handful of organic and natural meat brands starting in 2002, and sold them to Perdue Farms after nine years. The DOJ conducted a brief investigation of this acquisition, related to concerns about concentration in chicken processing but allowed it to go through (Department of Justice 2011b). By bundling competitors and selling them to larger firms, venture capitalists speed up the process of concentration. Many of the changes in ownership they catalyzed among organic brands would likely have occurred even faster if it had not been for the financial crisis of the late 2000s, although the pace of acquisitions began to increase again in 2013.

Table 8.1 Major North American independent organic firms and subsidiary brands

Alvarado Street Bakery	Equal Exchange	Sno Pac
Amy's Kitchen	Frontier Natural (Simply Organic)	Springfield Creamery (Nancy's)
Bob's Red Mill	Lundberg Family Farms	Traditional Medicinals
Cedarlane	Nature's Path (Envirokidz, Que Pasa)	Yogi Tea
Clif Bar (Luna)	Pacific Natural Foods	
Eden Foods	Organic Valley (Organic Prairie)	

the power to negotiate better terms with distributors, particularly when their products make up a significant percentage of the distributors' sales.

The decision to remain independent therefore does not make sense from a strict economic perspective. The founders of these firms have retained some of the idealism that motivated their entry into the organic food industry and do not want to see those ideals compromised. From their own experience or those of others in the industry, they have learned that acquisitions can have negative consequences for their non-economic goals. Greg Steltenpohl, formerly of Odwalla, has publicly expressed regret about losing financial control of his company, and the resulting emphasis on profit. He said, "[Corporations] have an agenda to consolidate and concentrate power and wealth. That's what their function is … The system itself forces certain outcomes, and I really underestimated that" (2005). This recognition is a stark contrast to the optimistic "we're not selling out, they're buying in" language founders have often used when announcing the sale of their firms (Rosenwald 2008).

Firms that have remained independent have also been more willing to publicly engage in the debates surrounding standards. Eden Foods, for example, does not use the USDA organic seal on its products, even though they are certified by the agency, because "this seal does not approach Eden's high standards for organic, in practice or in spirit" (2006). The company's chair and president, Michael Potter, has also been an outspoken critic of the OTA, stating: "Eden Foods has never been a member of the OTA despite urging from industry peers to join" and that "the OTA has worked consistently to weaken organic standards"

(Eden Foods 2005). Another independent, Equal Exchange, which is nearly 100 percent organic and entirely fair trade, has been vocal about changes in fair trade rules that allowed coffee plantations to be certified in 2012. The worker-owned cooperative withdrew from the certifying body, Fair Trade USA, and challenged the larger, publicly traded, Green Mountain Coffee to do the same via full-page newspaper advertisements (Howard and Jaffee 2013).

Conventionalization? Farming and ranching

The farming of organic foods, outside of a few product categories and regions of the country, has been more immune to the trend toward economic concentration when compared to the segments discussed above. One reason is that farming and ranching is much riskier than processing, distribution, or retailing. As discussed in Chapter 6, such factors as weather, pests, diseases, and long periods of time between purchasing inputs and selling agricultural products combine to make agriculture a less attractive place to invest money in the pursuit of higher profits (Mann and Dickinson 1978; Goodman, Sorj, and Wilkinson 1987; Kautsky 1988). Few farmers experience the temptations of lucrative buyouts faced by pioneering organic entrepreneurs in other parts of the organic food system. The additional restrictions imposed by organic standards present even more barriers to those with aspirations to scale up their operations (Buck, Getz, and Guthman 1997). Yet the market success of organic has attracted interest from larger producers, some of whom have sought to change the national standards to reduce these barriers.

One such effort was only temporarily successful. Fieldale Farms expressed interest in feeding as much as 20 percent non-organic grain to chickens while labeling the meat organic, in order to obtain a premium price for the meat without having to pay a premium themselves for inputs. The firm was unsuccessful in lobbying the USDA, so their next step was to approach a member of Congress from their district in Georgia, Nathan Deal. They convinced him to attach a rider to an appropriations bill to legalize the practice in 2003, but its passage was immediately challenged by a number of organizations and was repealed with a separate bill before it could go into effect. OTA Executive Director Katherine DiMatteo complained that: "It was a case of someone who didn't want to accept the regulations and the standards managing through the dark of the night and behind closed doors to actually get what they wanted" (Weinraub 2003)—as noted above, however, she and her organization learned from this experience and used the exact same strategy to allow synthetics to be used in processed organic foods a few years later.

Milk is one of the most popular organic products, and organic dairy production has become much more concentrated since the introduction of a national standard. The original USDA definition requiring "access to pasture" was vague, which was exploited by new and highly capitalized entrants. Horizon, owned by conventional dairy giant, Dean Foods, and Aurora, a private label dairy co-founded by natural/organic entrepreneur, Mark Retzloff, each had herds totaling more than 10,000 cows—the USDA eventually determined that these animals had inadequate access to outdoor grazing. Aurora was found to be willfully violating the organic standards in 2007, but the USDA allowed the firm a full year to achieve compliance while continuing to market their products as organic. The NOSB agreed to a more specific standard in 2008, requiring 120 days of access to pasture, although it took nearly two years for the USDA to implement the new rule. Horizon, Aurora and several other operations continue to run factory-scale dairies with thousands of cows, and the Cornucopia Institute charged in 2014 that Aurora was still out of compliance with the regulations (Fantle 2014).

A number of studies have explored Buck, Getz and Guthman's (1997) prediction that national standards and market success would lead organic farming to become increasingly similar to conventional farming. Reflecting trends in the food system in general, they expected organic farming—particularly for high value crops—to rely more on industrial equipment, contracts with processors, and vertical integration. In addition, because the national organic standards focused on the more easily enforceable prohibition of inputs rather than agroecological processes, they suggested capitalists would find allowed alternatives for products that had previously been produced on the farm, such as applying imported sodium nitrate for use as fertilizer instead of planting cover crops (Guthman 2004a).

These studies of "the conventionalization thesis" have had mixed results, however. In some regions, particularly California, where conventional agriculture is traditionally larger in scale and more capital intensive, these predictions have been more accurate (Guthman 2004b). Certain crops with production centered in this region, such as leafy greens and carrots, are also more likely to be large scale, as they are more amenable to mechanization. In other regions—including elsewhere in the United States, Canada, Australia, New Zealand, and Europe—and for other products, research has found less support for these changes (Campbell and Liepins 2001; Hall and Mogyorody 2001; Lockie and Halpin 2005; Best 2008; Constance, Choi, and Lyke-Ho-Gland 2008; Guptill 2009). The majority of studies investigating conventionalization have found some evidence of: (1) larger farm sizes, (2) a higher percentage of farms that include conventional practices on part of their operation, and (3) increased financial motivations for certification. On the other hand, they have also identified factors that limit these tendencies from going too far, and in some cases, encourage organic practitioners to become less reliant on agricultural inputs (Guptill 2009; Campbell and Rosin 2011).

From mom and pop to Monsanto: agricultural inputs

Organic food production is, in theory, supposed to rely primarily on farm-produced inputs rather than sourcing them off the farm. In practice, fruit and vegetable production is often separated from animal agriculture, thus requiring growers to seek off-farm sources of manure and other fertilizers. Some of the largest organic farms in California—Pavich, Cal-Organic, and Bornt & Sons—successfully pressured an influential California certifier to allow exceptions for sodium nitrate (also known as Chilean nitrate, named for the country where it is mined) rather than introduce a ban in accordance with numerous other certifiers (Guthman 2004a). The certifier compromised by limiting applications of sodium nitrate to 20 percent of crop nitrogen needs. This specific cutoff was adopted in the USDA's national standard, but in confirmation of the conventionalization thesis, it was relaxed to allow unrestricted use in 2012.

Organic fertilizers have become a much larger industry since the introduction of the national standard. The potential for high profits led at least two firms, California Liquid Fertilizer and Port Organic Products, to engage in fraudulent behavior, such as adding large amounts of synthetic fertilizer to products labeled as approved for organic production. The owners of both firms were sentenced to prison terms for their actions. The farmers who purchased these fertilizers, however, were not penalized by certifiers or regulators, and their identities were kept secret. In contrast, in cases where synthetic pesticides were mistakenly applied to crops, farmers typically have had their certification suspended for three years. The disparity was explained by the USDA as due to the farmers' inability to detect the fraud and because "the affected products posed no threat to the health of consumers or the environment" (USDA 2014b).

Controversies have also erupted over the allowance of specific pesticides to control pests or diseases. Some of these have included *Bacillus thuringiensis*, rotenone, pyrethrum, sulfur, copper, and some antibiotics. The philosophy of organic is to minimize these types of inputs and more idealistic organic farmers, if they use them at all, do so only as a last resort. Less idealistic growers, such as some large-scale grape growers, lobbied organic certifiers to allow these toxic chemicals, despite their negative impacts on workers (Guthman 2004a). In some contexts it is more difficult to avoid applications of approved organic pesticides, as in the case of wetter climates, which are more conducive to fruit tree diseases. Most of these substances originally approved by the USDA standards are still allowed, although strong opposition to the application of the antibiotics, streptomycin and oxytetracycline, on apples and pears, to control bacterial fire blight, contributed to the NOSB decision to allow approval to expire in 2014.

Seeds are another key input for many organic farmers. The labor and knowledge-intensive nature of seed saving, particularly with a diverse farming operation, often leads to seed purchases from off-farm suppliers. As discussed in the previous chapter, this industry has become dominated by chemical giants like Monsanto and DuPont, even for seeds commonly used by organic farmers and gardeners: examples include Big Beef and Early Girl tomatoes, Stars "n" Stripes watermelon, and Early Butternut squash (Dillon 2005). Alfonso Romo Garza, a Mexican billionaire, observed the consolidation that was occurring in the commodity seed industry in the early 1990s and saw an opportunity to do the same for fruit and vegetable seeds. He formed a conglomerate called Seminis and by 1998 had gained control of 39 percent of the market for vegetable seeds in the United States, 24 percent in the EU, and 26 percent globally (Dillon 2008). He sold this firm to Monsanto for $1.4 billion in 2005.

Given Monsanto's primary role in commercializing GE organisms, members of the organic community were understandably upset about the prospect of financially supporting this firm. Some seed retailers that catered to organic farmers and gardeners, such as Fedco, dropped Monsanto/Seminis varieties completely. Others, such as Territorial and Johnny's, labeled the source and continued to sell them until they were able to phase them out. Individual farmers also took actions to regain self-reliance, such as a Santa Cruz, California farmer who developed an open-pollinated version of the Early Girl tomato, in contrast to Monsanto's hybrid variety (Duggan 2014).

The national organic standards were lax in requiring farmers to source certified organic seed at first, but they have since been somewhat tightened. Initially, concerns about inadequate supplies of certified organic seed, particularly at prices that were comparable to conventional seed, led certifiers to allow numerous exceptions. Pressure to decrease the industry's reliance on conventional seeds motivated increased paperwork requirements for organic growers, who must now demonstrate "good faith" efforts to confirm the unavailability of organic seed (Renaud, Bueren, and Jiggins 2014). The rule changes have contributed to the success of organic seed companies such as Seeds of Change (Box 8.3) and High Mowing Seeds of Vermont. The latter has grown exponentially since being founded in 1996.

Box 8.3 Maintaining Secrecy: Seeds of Change

Seeds of Change, a 100 percent organic seed company, was acquired by M&M Mars in 1997. At first glance the purchase appears puzzling—why would a company focused on mass market candy buy such an unrelated business? M&M Mars is one of the largest privately held firms in the world, which allows it to be one of the most secretive, so definitive answers are hard to find. The president of Seeds of Change at the time, however, was twenty-eight-year-old Stephen Badger, a grandson of the late Forrest Mars, Sr.—his mother and two uncles are estimated to be three of the forty wealthiest people on the planet (Dolan and Kroll 2014). Badger later became a member of the family-controlled board of M&M Mars. Shortly after the ownership change, Seeds of Change expanded into a slightly more related business of processed organic foods, such as pasta sauces and salad dressings. These products were sold in the United States, as well as the UK, Ireland, Scandanavia, and Australia. The new parent company was said to be relatively hands off at first, and Seeds of Change subsequently established two seed breeding facilities in Scandanavia. More recently, however, the original farm in El Guique, New Mexico was shut down and some employees were transferred to Los Angeles (Dyer 2010).

Seeds of Change was started in 1989 by Gabriel Howearth, Kenny Ausubel (who later co-founded the Bioneers conference), Emigdio Ballon, and Alan Kapuler. An investor in the company, former university professor Howard-Yana Shapiro, replaced Ausubel in 1995 (Waters 1996). Shapiro played a key role in the M&M Mars buyout and remains with the firm as their chief agricultural officer. After the transition, Seeds of Change was scored poorly on sustainability criteria by the UK magazine Ethical Consumer, but Shapiro responded dismissively: "We're a private company, it would be hard for them to know anything about us" (Jowit 2009).

Beyond organic?

Advocates of the less industrial vision of organic have used numerous strategies to resist changes to the meaning of the term. In 2002, for example, the pioneering California organic farmers Rick and Kristie Knoll decided to give up organic certification, just as the national standards went into effect. They were frustrated both with federal government/corporate involvement and, like Tuttle in Brazil, the increasing amount of paperwork required for compliance. Rick bought a new surfboard, anticipating a big drop in sales that would free up more time for him to ride the waves. Years later, although some retailers stopped carrying their products, he still had not used his surfboard! The positive reputation the Knolls had built with the customers at restaurants and farmers' markets led to increasing sales and little time for non-farming pursuits (Harper 2005).

Although direct markets account for a very small percentage of organic food sales, such outlets play a crucial role in supporting small-scale farmers. These types of farmers are more likely to have very diverse operations and have often been important sources of innovation, such as developing season extension systems. Those like the Knolls, who have abandoned certification, are more likely than certified farmers to report that their customers trust them (Constance, Choi, and Lyke-Ho-Gland 2008). Producers that sell outside of direct markets may not have the same level of trust, and therefore face demands from distributors and retailers to verify their claims. Typically, these buyers prefer that a third party, with no direct financial interest in the outcome, be responsible for certifying compliance with standards, as is the case with USDA organic.

One response from producers and their allies has been to organize other types of ecolabels, with standards that differ from organic (Table 8.2). Some of these are intended to complement the existing USDA organic standards and highlight criteria that are no longer embodied in them, while others go "beyond

Table 8.2 US ecolabels that complement or go "beyond" organic

Multiple criteria	Humane
Demeter Biodynamic	Animal Welfare Approved
Certified Naturally Grown	Certified Humane Raised and Handled
	American Humane Association
Social justice	
Food Justice Certified	**Animal feed requirements**
UFW (United Farm Workers) Black Eagle	American Grassfed
	USDA Certified Grass-Fed
Wildlife protections	
Salmon-Safe	**Geographic origin**
Predator Friendly	Buy Fresh Buy Local

organic" and set a higher bar (Howard and Allen 2010). Potentially complementary third-party certified ecolabels include several competing humane treatment of animal certifications and grass-fed certifications; some of these offer the advantage of just one site visit to obtain dual certification with USDA organic.

Biodynamic is not a new label, but farms using this system may now receive dual certification with USDA organic via the Demeter Association. Biodynamic requirements follow philosophical principles developed by Rudolf Steiner in the early 1900s and encompass multiple criteria, such as a greater emphasis on the production of inputs within the confines of the farm. Some consumers seek out biodynamic because the bar is higher than organic, but others are drawn not by philosophical or even ecological reasons but for taste. Many California wineries have adopted biodynamic practices for this reason, as they explain on their websites, but they are more reluctant to mention it on the product label (e.g., Quivira 2014). Although several hundred US farms are Demeter certified, these are just a small fraction of the estimated 6,000 farms using biodynamic practices (Karp and Carlson 2014).

Certified Naturally Grown is another multi-criteria label, and it was developed to provide an alternative to some of the common farmer complaints with organic certification. The standards are based on those of the USDA, with some additional "beyond organic" requirements, but certification costs less, and it replaces a large stack of paperwork with a simple declaration. Instead of the third-party system, participating farmers are asked to certify their neighboring farms. Retailers, including food cooperatives, are more skeptical of this "peer-certification" process, but committed organic consumers report placing more trust in Certified Naturally Grown than USDA organic (Spaniolo and Howard 2010). More than 700 farms participated in the former by 2014 (Certified Naturally Grown 2014).

All of these efforts are quite small, however, in comparison to the 18,513 certified organic farms and processing facilities in the United States identified at the end of 2013 (McEvoy 2014). As this chapter has detailed, the rapid growth of the organic industry has attracted involvement from dominant firms, particularly in the processing stage of the food system. This trend has resulted in a disproportionate influence on the "rules of the game," which have weakened processing standards more than production or input standards, despite organized resistance from more idealistic members of the organic community.

The results have not been entirely negative, however. Even the largest organic processors must source from certified organic farms, which has reduced the amount of synthetic pesticides and other unapproved inputs applied to farmland.

Although transnational firms have been dismissed as "dabbling" in organic for just a small percentage of their operations, practices to reduce reliance on expensive inputs have also been adopted by an increasing number of conventional producers. The benefits to ecosystems and human health from these changes should certainly not be underestimated—even if the primary motivations may have been financial or ultimately with a goal of increasing power. The following chapter discusses likelihood of reducing even more of the negative impacts of dominant firms, in the broader context of exploring potential future outcomes for the food system.

Chapter 9

Endgame?

You will be assimilated. Resistance is futile.
—The Borg (Star Trek: The Next Generation)

In the final stages of a chess match—the endgame—there are typically just a few pieces left on the board. One possible future resulting from the trend of increasing market concentration is described by a voice on the street in the satirical newspaper *The Onion*; in response to a recent buyout announcement, a woman says, "it's kind of comforting to know everything will be owned by one or two people someday" (2013). Will just a few firms end up controlling the entire world food system, from supermarket shelves to seeds—in other words, an endgame of global monopoly? The elite individuals who benefit from these trends would suggest the answer is yes and push their own version of The Borg's admonition that "resistance is futile." The prime minister of the UK in the 1980s, Margaret Thatcher, for example, famously said, "There is no alternative" to the capitalist system, which is sometimes shortened to the acronym TINA (George 1999).

I expect few food industries will become global monopolies, but this is a small consolation, as more sectors are approaching domination by just two firms. As previous chapters described, however, firms' strategies to shape and reshape society are almost always resisted—although this resistance may be quite hidden. Most of these counter-movements have failed to slow consolidation, but some have managed to address key negative impacts, despite the relatively small numbers of people involved. Others have demonstrated that alternatives *are* possible, even in a system organized to provide the greatest advantages to large, capitalist firms. This ongoing resistance also illuminates some potential limits to increasing power, as well as negative feedback loops that could quickly halt or reverse current levels of concentration. It is difficult to determine which of these feedbacks is most likely to have such an impact, but future research could assist in identifying the most promising paths for creating a more decentralized, socially just and ecologically sustainable food system.

Trends toward global duopolies

Although near monopolies in some countries are possible, such as the 78 percent share of the beer market held by SABMiller in South Africa (Ascher 2012), this outcome is less likely at the global level. Duopolies, such as Coke and Pepsi's dominance of the soft drink industry, offer more advantages for both firms and governments. This model provides an appearance of competition, yet still allows for stable monopoly profits. It may even increase sales, as an "enemy" firm or brand may exploit the psychology of social identity to increase consumer loyalty (Dooley 2011). While monopolies may be tolerated for long periods in some industries, such as computer operating systems or microprocessors, food is typically too essential to allow just one firm to control an entire market—such a development could seriously undermine the legitimacy of the government.

The beer industry is already a duopoly in many countries and is fast approaching this state at the global level. Wall Street investors are fond of speculating that AB InBev will take over the second-ranked firm, SABMiller, because the third and fourth-ranked firms are deemed not valuable enough to acquire; in 2014, AB InBev reportedly explored financing such a move but did not complete a deal. The governments of China and the United States would be very unlikely to approve such a combination, however (Brown 2014). With this possibility blocked, an acquisition of Heineken or Carlsberg might become more attractive; Heineken turned down a proposal from SABMiller in 2014.

The seed industry is also nearing domination by just two firms. In early 2015, third-ranked Syngenta rejected an unsolicited $45-billion buyout offer from the top firm, Monsanto, but analysts expect a higher bid in the near future. For name brand seeds, the choices are increasingly limited to Monsanto and DuPont's offerings, whether buyers know this or not (see Box 9.1). Governments have

---

Box 9.1 Uncovering Stealth Ownership

Collective efforts to shift purchases away from dominant firms represent one strategy to oppose industry consolidation but one that requires up to date knowledge of where our money really flows. There is no single way to determine which firm owns a product because most countries do not require disclosure of this information to consumers. There are, however, several methods you can use to uncover clues or perhaps even solve the mystery of who owns what.

A good first place to start is one of two websites/apps that have ownership information, GoodGuide.com and Buycott.com. Both allow you to take a picture of a bar code to pull up information about that product or to type the product name into a search engine. GoodGuide is not always as up to date as Buycott for ownership information but does provide additional information, including scores for health, society and environment compared to other products in the category.

The next place to try is the brand's website, as sometimes the corporate parent can be found somewhere in the fine print. Smucker's stealth ownership of its organic juice brands is revealed if you check a tiny link to their "Transparency in the Supply Chain" page. Thanks to a California law that went into effect in 2012, firms that operate in the state and gross over $100 million are required to disclose their efforts to eradicate slavery and human trafficking by their suppliers. Some firms find it easier to comply with this law by using a boilerplate document for all of their brands. Also, be sure to notice how professional the website looks. An extremely slick design can be a clue that the company is part of a very large corporation and encourage you to keep searching for more information.

Another source of information is a trademark database. The US Patent and Trademark Office provides a free, searchable database, but there are others that are more user friendly, such as trademarks.justia.com. There is a good chance that this strategy will reveal the corporate parent, but it will not work in every case. Some firms will keep the trademark in the subsidiary's name or use the names of outside attorneys for this paperwork. Look for the most recent trademark (with the highest serial number) because sometimes you can observe multiple ownership changes through these documents.

Finally, some web searches that use the name of the brand along with keywords like "acquisition" or "equity stake" may turn up articles in business or trade publications that announce buyouts. Conversely, if you have a potential parent firm in mind, you can search the firm's website, press releases, annual reports, and Securities and Exchange Commission (SEC) filings for brands they have acquired. While the parties involved may not want ordinary customers to know about these ownership changes, they typically want investors to be informed about efforts to increase their power.

Examining ownership in food industries that do not have well-known brand names, such as ingredient suppliers and distributors, can be even more challenging. Don't be discouraged if you find it difficult to uncover who really owns everything that you buy. Even people working in these industries can't always keep track of the constant mergers, acquisitions, changing equity stakes, and strategic alliances. I am always surprised when CEOs, heads of marketing departments, and market researchers tell me that they find my visualizations of industry structure to be useful.

been more reluctant to take antitrust action in this industry, perhaps because it is not as visible to most of the public when compared to beer.

For many other food industries, such as supermarkets, trends toward duopolies are occurring more slowly. Walmart, Tesco, and other global giants have failed to sustain the growth rates achieved in the 1990s, but attempts to resolve this dilemma by merging would likely lead to more scrutiny from regulators. Walmart has not made major buyouts in the United States, like it has in other parts of the world. This was not for lack of interest. Walmart repeatedly proposed to acquire Meijer, a Michigan-headquartered chain of supercenters, from the 1970s to 1990s, but the owners refused to sell (Crawley 2009). Their resistance significantly slowed Walmart's supercenter growth in the upper Midwest during this period. Walmart's currently high market share would likely prevent the firm from acquiring even a regional chain with US government approval, despite numerous national competitors, including Kroger, Safeway/Albertson's, Target and Costco.

Whether or not their industries are approaching global duopolies, dominant firms are also likely to attempt to expand in additional directions, such as concentrically or vertically. Concentrically, they may diversify into related industries, as with supermarkets moving into the convenience store format or the agricultural chemical industry's takeover of much of the seed industry. Another possibility is for beer and soft drink firms to integrate further, such as converting an existing alliance between AB InBev and Pepsi into a full acquisition of the latter.

Vertically, some firms have successfully dismantled previous barriers to concentration in upstream or downstream segments of their supply chains, using approaches that have included influencing government policies, developing new technologies, increasing control over contract growers, and forming alliances with other dominant firms. AB InBev's success in changing three-tier alcohol policies, for example, allowed the corporation to increase its ownership of the distribution stage of the beer supply chain. New technologies have enabled global pork processors to become more directly involved in the production stage, but these firms also minimize their risks by sourcing from thousands of additional hog farmers, who must adhere to very one-sided contracts. The alliance strategy is illustrated by DuPont, which cooperated with a grain trader, Bunge, and a packaged food manufacturer, General Mills, to develop a value-added soybean supply chain. Under the Solae name, this chain extends from ingredients in branded foods and beverages—such as 8th Continent brand soy milk—to DuPont's seeds, with operations on four continents.

The increasingly global scope of many food and agricultural firms has reduced the power of national governments to control their behavior, which gives them additional advantages over smaller firms. The fast food chain Burger King, for example, recently "merged" with a Canadian coffee chain, Tim Horton's; through a tactic called "inversion," the headquarters of the combined firm was located in Canada to lower its tax payments. Because transnational firms can pit governments against each other, numerous US legislators responded to the deal with calls to further reduce the corporate tax rate (Isidore and Sahadi 2014). It is also likely that antitrust enforcement will be weakened further in response to such political pressures. Not all sectors of society are acceding to the demands of dominant firms, however, and a diverse range of resistance tactics are being employed by counter-movements.

Counter-movements

Resistance to dominant food and agricultural firms' efforts to increase their power may have changed some of the strategies they use but, as noted above, failed to reverse trends toward increasing their market share. Challenges to these firms' negative impacts on human health, the environment, animal welfare, and labor practices, however, have experienced some successes. Campaigns to remove objectionable ingredients from packaged foods discussed in Chapter 4, for example, have also affected commodity processors. Perhaps the most high profile case involved "pink slime," known to the industry as lean, finely textured beef. This product results from heating, centrifuging, extruding, and applying ammonia to meat that might otherwise be discarded, and it was added to at least half of the ground beef and hamburgers in the United States. In 2012, extensive media coverage contributed to successful consumer demands to remove pink slime from school cafeterias and fast food restaurants. As a result, one manufacturer declared bankruptcy and another closed several plants where it was made (Velasco 2012).

A less publicized case was the dramatic decline in the use of rBGH, due to human health and animal welfare concerns. Rick North of Oregon Physicians for Social Responsibility led a letter writing campaign targeted specific processors and retailers, asking them to stop buying milk from farms that use the product. Slowly and steadily, the campaign pressured a number of leading firms to agree to this demand, reducing its use in the United States from an estimated 90 percent to 25 percent (Dalton 2011). Sales of the product declined to the point that Monsanto sold the rights to Eli Lilly in 2008. This outcome was especially impressive when considering that FDA guidelines discouraged rBGH-free labels. These were written by a former Monsanto lawyer Michael Taylor, who later became a Monsanto vice president and then a deputy commissioner at the FDA; he is frequently used as an example of the "revolving door" between government and industry (Mattera 2004).

Meat processors have also responded to public pressure to end the use of hog gestation crates, which typically measure seven feet long and two feet wide. Animal scientist Temple Grandin has described them as inhumane, asking: "How would you like to live in an airline seat?" (Carman 2012). Legislation has required gestation crates to be phased out in New Zealand, Australia, a number of EU member states, and several US states. In addition, Smithfield agreed to phase them out entirely by 2022. Smithfield also dropped the feed supplement ractopamine from approximately half of its American hog operations; its use to promote lean muscle growth is common in the United States but banned in nearly all other countries (Huffstutter and Baertlein 2013).

Positive changes in the tomato industry have resulted from public pressure on retailers, which contract with growers to buy these products. Tomato farming is highly concentrated on large operations in California and Florida. In response to organizing by the Coalition of Immokalee Workers, a dozen retailers have agreed to pay an additional penny-per-pound for fresh tomatoes from Florida, beginning with Taco Bell in 2005. This amount results in a 20–35 percent increase in pay for the workers and has improved workplace safety (Greenhouse 2014).

These and many other efforts directed at dominant firms have succeeded in steering "business as usual" toward small but positive changes. Similar impacts have resulted from capitalists' incorporation of alternatives to the mainstream food system (e.g., craft beer, organic, fair trade, local produce), although typically at very low levels (Jaffee and Howard 2010; Howard and Jaffee 2013). While these outcomes may also result in watering down or co-opting some of the original ideals, they demonstrate public support for alternatives to the status quo, as well as potential avenues to create desired changes.

There are numerous other alternatives that are either being ignored or more directly undermined by dominant firms. Many of these are nearly invisible, because they are not part of the formal economy. As just one example, nearly half of the global milk supply bypasses industrial processing altogether and is consumed raw or as yogurt, fermented drinks, or cheese. Capitalists have sought to capture these potential markets but have experienced some important failures. In 2006, Colombia passed a law modeled on more industrialized countries that prohibited the consumption, sale, and transport of unpasteurized milk, but widespread protests resulted in years of postponements and eventually its repeal (GRAIN 2012).

Dominant trends toward concentration result in a mass market that is increasingly under the control of fewer people. At the same time, counter-movements that directly or indirectly oppose them result in niches that place power in the hands of more people—although with stealth ownership it is sometimes challenging to distinguish between the two. The relative market share for such niche segments differs depending upon the level of development reached by the food system and the specific industry, and most alternatives in industrialized countries remain quite small. The US beer industry provides one of the clearest illustrations of these trends: the dwindling number of dominant firms have a relatively flat market share, while a growing number of craft brewers have increased their collective share to 14 percent of sales. This is what motivated investors to express interest in an AB InBev-SABMiller tie-up, as they do not see many other avenues for these firms to boost stock prices. Examples like these suggest that some food and beverage industries may be approaching more insurmountable limits to increasing power.

Box 9.2 Remaining Questions

The questions raised at the beginning of this book—how is concentration changing in the food system, and what factors enable or constrain the goals of dominant firms—have certainly not been fully answered. While I have outlined some broad quantitative trends, primarily in the United States, and explored some of the key qualitative strategies of dominant firms, much remains to be discovered. Obtaining sales data for entire industries remains a substantial challenge, which makes it even more difficult to disentangle concentration from other trends (e.g., industrialization) or to accurately determine its impacts on communities and ecosystems. Nevertheless, future research on concentration and power in the food system would benefit from a better characterization of dominant firm market shares in more industries and nations, as well as historically and globally.

Concentration studies and value chain analyses are beginning to borrow more tools from social network analysis. A network analysis of the ownership of transnational corporations, for example, reported that control is distributed extremely inequitably and that for the most densely connected network, fewer than 1 percent of the companies were able to exercise control over 40 percent of the wealth—the most dominant firms were mostly financial institutions, such as Goldman Sachs and Barclays (Coghlan and MacKenzie 2011; Vitali, Glattfelder, and Battiston 2011). These methods also open up possibilities to better explore how power is exercised without direct ownership, such as through the structure of contingent exchanges, contracts, and strategic alliances. Mapping the relationships between key individuals in dominant firms and those in other institutions, such as governments, foundations, universities and think tanks, would assist with understanding the qualitative strategies that are employed to increase their quantitative market shares.

Within the Capital as Power framework, more research is needed on the perspectives of dominant capitalists, such as how they compare themselves to other firms, both within their industry and outside of it, as well as what motivates them to engage in more cooperative or competitive behaviors with these firms. Content analysis is unlikely to fully answer these questions, and ethnographies of corporate executives are difficult given the restrictions on academic research involving human subjects. Journalists and activists might be better positioned to conduct such investigations. Brewster Kneen, for example, has described how he was able to simply walk into the offices of the private firm Cargill and talk with their employees, as well as collect some of their internal newsletters (2013).

More approaches to link quantitative assessments of power and the influence of qualitative strategies on these measures are also essential. This is complicated by the fact that historical market capitalization data is expensive for researchers to access. Prospective studies offer a potential way around this difficulty by recording the more accessible current data on an ongoing basis. One possibility is to explore the growing use of tournaments to determine contractors' compensation in food and agricultural industries and its effects on firm power. Tournaments are already prevalent in some sectors in which it is difficult for capitalists to raise concentration levels, such as beer distribution, chicken production, and corn seed production. As tournaments are introduced to additional sectors, it would be interesting to analyze the impact on firms' market capitalization, particularly if it does not influence their market share.

Capital as Power would also benefit from integration with other frameworks, such as political ecology, actor-network approaches, and social movement theories. Both political ecology and actor-network approaches, for example, attribute more power or agency to crops, livestock, and other nonhuman components of agroecosystems and critically assess their relationships with human actors (Galt 2013). Social movement perspectives on Capital as Power are also needed to better theorize resistance and the contexts in which it is most or least effective (Bousfield 2013).

Throughout this book, I have emphasized how concentration is hidden in many industries. Another important research need is to identify the best ways to communicate these trends and their impacts to the public, so that people can make more informed decisions in civic engagements with these issues. I have found visualization to be one effective means of quickly communicating complex changes, but less is known about the relative advantages and

> disadvantages of specific formats, such as treemaps, cartographic maps, and cluster diagrams (Howard 2009a). Many other forms of community engagement also need to be studied for their effectiveness in motivating collective actions, particularly less top-down, more participatory methods (Howard 2011).

Potential limits?

The food system has to this point provided a number of positive feedback loops for dominant firms—as they increase their power they accrue even more advantages, which then reinforce their ability to restructure society. These positive feedbacks are unlikely to continue indefinitely and some may already be approaching limits or asymptotes that threaten to undermine the stability of the entire system (Bichler and Nitzan 2012). As firms move closer to these limits, the negative feedbacks increase, which require more force to counter, further accelerating the feedbacks. Social resistance is one such feedback, and numerous examples have been described in the previous pages. Frequently, just a small minority of the population is actively participating in movements to oppose dominant firms, but as the case of pink slime demonstrated, efforts can suddenly resonate with much larger numbers of people, leading to rapid changes in the system.

Executives at dominant firms have occasionally contributed to increasing resistance by publicly expressing their views in terms that were perhaps too blunt. Some examples include:

> There isn't one grain of anything in the world that is sold in a free market. Not one! The only place you see a free market is in the speeches of politicians. People who are not in the Midwest do not understand that this is a socialist country.
>
> —Dwayne Andreas, CEO of Archer Daniels Midland (Carney 1995)

> The one opinion, which I think is extreme, is represented by the NGOs, who bang on about declaring water a public right. That means as a human being you should have a right to water. That's an extreme solution.
>
> —Peter Brabeck-Letmathe, CEO of Nestlé (Wagenhofer 2005)

> In the broadest sense, Cargill is engaged in the commercialization of photosynthesis.
>
> —Greg Page, CEO of Cargill (Page 2008)

Such ill-advised statements help to illustrate the differences between their calculated public images and the far less altruistic strategies in which they actually engage away from public view.

One mechanism for ensuring social acceptance is exercising the power to "*suppress awareness* of the preponderance and root causes of inequality and injustice" (Fridell and Konings 2013, 7). This awareness can be suppressed in numerous ways, such as using public relations firms to script the basis of more than half of all mass media news stories (Bacon 2010), providing "educational" materials to public schools (Schor 2004) and controlling oppositional movements through foundation funding (Smith 2009). A common strategy involves "diversionary reframing," which is another way of saying that the powerful attempt to change the subject and shift attention to less controversial topics (Freudenberg and Alario 2007). One example in food industries is the "feeding the world" rhetoric of grain traders discussed in Chapter 5; this is also very popular with seed-chemical firms to legitimize transgenic organisms, despite little evidence to indicate they have increased productivity (Peekhaus 2013). Another is the appeal to "cheap food" by nearly every sector, which directs attention away from numerous social and ecological costs embodied in food production (Carolan 2011). As the power of dominant firms increases, however, it becomes more difficult to hide their influence on other organizations and society at large.

Public awareness of income and wealth inequality has increased dramatically since the 2008 financial crisis, aided by the Occupy movement protests against the richest 1 percent of the population in 2011. This crisis also resulted in greater exposure of the role of governments in maintaining these inequalities, such as the billions of dollars spent on bailouts for financial institutions and mass surveillance programs that handed over foreign business information to US-headquartered firms (Greenwald 2014). These trends are beginning to lower the market capitalization of many dominant firms, suggesting that although profits as a share of income are higher than ever, investors' expectations that the public will continue to acquiesce to such a level of control are weakening (Bichler and Nitzan 2012).

This is particularly true of the food industry, because the poorest members of society are more sensitive to price changes in this sector when compared to most other sectors of the economy. Recent studies suggest that food price increases are associated with social unrest, such as demonstrations and riots, especially in less industrialized nations (Arezki and Brückner 2011; Bellemare 2014). Prices for food increased rapidly in many countries between 2006 and 2008, for example, and led to riots in Africa, Asia, Latin America, and the Middle East. The expanding emphasis of food and agriculture firms on financialization was likely a key driver of this inflation, and the resulting protests contributed to the overthrow of governments in Haiti and Madagascar (Goodman, DuPuis, and Goodman 2012).

In addition to these social limits, there are potential natural limits to trends in the food system, such as dwindling supplies of fossil fuels, key fertilizers, pollinating species, water, and fertile soil, as well as increased possibilities of climate change, severe weather, pests, and disease (Barker 2007). As one illustration, California has experienced a drought since 2011, which may be its worst in 1,200 years. It threatens the state's heavily irrigated agricultural system, which produces approximately half of the fruits, vegetables and nuts consumed in the United States. Another is an outbreak of avian influenza in the Midwestern poultry industry, which in early 2015 resulted in the death or euthanasia of more than 30 million turkeys and chickens (McNeil 2015). Governments and dominant firms typically advocate technical fixes to these problems, because they increase, rather than decrease, their power. Yet this approach also tends to reinforce the very problems it is supposed to solve, bringing us closer to potentially catastrophic outcomes (DeLind and Howard 2008; Weis 2010).

Negative feedbacks from natural and social systems can reinforce each other, as well. The industrial food system's reliance on just-in-time delivery (e.g., Walmart's cross-docking) makes it quite vulnerable to disruption. Both natural disasters and labor movement strikes have demonstrated how quickly food supplies can run out for the majority of the local population, and the social unrest that may result. The UK, for instance, is heavily dependent on imports of food and energy and would face rapidly escalating costs if transportation networks were disrupted for more than one week (Harvey 2012).

The trend toward greater levels of concentration appears to be unabated, but there are signs that it may eventually fall victim to its own success. Due to the extreme complexity of food systems, it is nearly impossible to predict which negative feedback is likely to play the greatest role in this reversal. Although research might help identify some promising avenues on which to focus our efforts (Box 9.2), it is also important to maintain a wide array of strategies. Counter-movements play a key role in demonstrating alternative possibilities of food provisioning, preserving seed and livestock diversity, and maintaining the skills and knowledge that will be needed to replace the current system. Joining these movements and supporting the alternatives created by others could therefore be essential to maintaining our ability to feed ourselves in the future.

References

Aalberts, Robert J., and Marianne M. Jennings. 1999. "The Ethics of Slotting: Is This Bribery, Facilitation Marketing or Just Plain Competition?" *Journal of Business Ethics* 20 (3): 207–15.

Ackerman-Leist, Philip. 2013. *Rebuilding the Foodshed: How to Create Local, Sustainable, and Secure Food Systems*. White River Junction, VT: Chelsea Green Publishing.

Adams, Walter, and James W. Brock. 2004. *The Bigness Complex: Industry, Labor, and Government in the American Economy*. 2nd ed. Redwood City, CA: Stanford University Press.

Adamy, Janet. 2005. "Behind a Food Giant's Success: An Unlikely Soy-Milk Alliance." *Wall Street Journal*, February 1. http://on.wsj.com/1nBxpQ3.

Aistara, Guntra A. 2011. "Seeds of Kin, Kin of Seeds: The Commodification of Organic Seeds and Social Relations in Costa Rica and Latvia." *Ethnography* 12 (4): 490–517.

Alaniz, Maria Luisa, and Chris Wilkes. 1998. "Pro-Drinking Messages and Message Environments for Young Adults: The Case of Alcohol Industry Advertising in African American, Latino, and Native American Communities." *Journal of Public Health Policy* 19 (4): 447–72.

Alders, Robyn, Joseph Adongo Awuni, Brigitte Bagnol, Penny Farrell, and Nicolene de Haan. 2013. "Impact of Avian Influenza on Village Poultry Production Globally." *EcoHealth* 11 (1): 63–72.

Alexander, Eleanore, Derek Yach, and George A. Mensah. 2011. "Major Multinational Food and Beverage Companies and Informal Sector Contributions to Global Food Consumption: Implications for Nutrition Policy." *Globalization and Health* 7 (1): 26.

Allegretto, Sylvia. 2012. "The Wrecking Ball." *The Berkeley Blog*, http://blogs.berkeley.edu/2012/07/16/the-wrecking-ball-2/.

Allen, Patricia, and Martin Kovach. 2000. "The Capitalist Composition of Organic: The Potential of Markets in Fulfilling the Promise of Organic Agriculture." *Agriculture and Human Values* 17 (3): 221–32.

Allen, Patricia, and Alice Brooke Wilson. 2008. "Agrifood Inequalities: Globalization and Localization." *Development* 51 (4): 534–40.

Americans for Tax Fairness. 2014. *Walmart on Tax Day: How Taxpayers Subsidize America's Biggest Employer and Richest Family*. Washington, DC: Americans for Tax Fairness.

Andrews, David. 2012. "Antitrust Efforts Have Gone in Dustbin of History." *National Catholic Reporter*, March 6. http://ncronline.org/news/politics/antitrust-efforts-have-gone-dustbin-history.

Aoki, Keith, and Kennedy Luvai. 2007. "Seed Wars: Controversies Over Access to and Control of Plant Genetic Resources." In *Intellectual Property and Information Wealth: Patents and Trade Secrets*, edited by Peter K. Yu, 249–81. Westport, CT: Greenwood Publishing Group.

Arezki, Rabah, and Markus Brückner. 2011. *Food Prices and Political Instability*. Washington, DC: International Monetary Fund.

Aron, Nan, Barbara Moulton, and Chris Owens. 1994. "Judicial Seminars: Economics, Academia, and Corporate Money in America." *Antitrust Law & Economics Review* 25 (2): 1–33.

Asche, Frank. 2008. "Farming the Sea." *Marine Resource Economics* 23 (4): 527–47.

Ascher, Bernard. 2012. *Global Beer: The Road to Monopoly*. Washington, DC: American Antitrust Institute.

Ashworth, Will. 2012. "The Problem with Yum! Brands Push into China." *InvestorPlace*, May 10. http://investorplace.com/2012/05/the-dark-side-of-yum-brands-china-strategy/.

Associated Press. 2007. "EU Fines Brewers $370M in Price-Fixing Probe." http://nbcnews.to/1GDHVtH.

Associated Press. 2008. "Beer Heiress Could Be Next First Lady," April 3. http://nbcnews.to/1ArxN56.

Associated Press. 2014. "China Supplier Sold McDonald's, KFC Expired Meat," July 21. http://usat.ly/WqaTh7

Atlas, Terry. 1989. "Soybean Contract Expires, But Furor Doesn't." *Chicago Tribune*, July 21. http://trib.in/1DQrLMZ.

Austin, S. Bryn, Steven J. Melly, Brisa N. Sanchez, Aarti Patel, Stephen Buka, and Steven L. Gortmaker. 2005. "Clustering of Fast-Food Restaurants Around Schools: A Novel Application of Spatial Statistics to the Study of Food Environments." *American Journal of Public Health* 95(9): 1575–81.

Bacher, Dan. 2004. "The Water Kings of California." *Counterpunch*, December 30. http://www.counterpunch.org/2004/12/30/the-water-kings-of-california/.

Bacon, Wendy. 2010. "Over Half Your News Is Spin." *Crikey*, March 15. http://www.crikey.com.au/2010/03/15/over-half-your-news-is-spin/.

Baertlein, Lisa. 2013. "Yum's China Woes Slam Sales and Profits." *Reuters*, February 4. http://reut.rs/YK1X7L.

Baines, Joseph. 2014a. "Wal-Mart's Power Trajectory: A Contribution to the Political Economy of the Firm." *Review of Capital as Power* 1 (1): 79–109.

Baines, Joseph. 2014b. "Food Price Inflation as Redistribution: Towards a New Analysis of Corporate Power in the World Food System." *New Political Economy* 19 (1): 79–112.

Baines, Joseph. 2015. *Price and Income Dynamics in the Agri-Food System: A Disaggregate Perspective*. Toronto, ON, Canada: York University.

Balu, Rekha. 1998. "Anheuser-Busch Amphibian Ads Called Cold-Blooded by Doctors." *Wall Street Journal*, April 10. http://on.wsj.com/1hrdfAp.

Bamber, Andy. 2014. "Annie's Homegrown: From $1.3 Million DPO to $820 Million Buyout." *Cutting Edge Capital*. http://www.cuttingedgecapital.com/annies-homegrown-dpo-to-buyout/.

Banjo, Shelly. 2012. "At Wal-Mart, Maturity Means Fat Dividends." *Wall Street Journal*, June 1. http://on.wsj.com/1qXeDmj.

Barabási, Albert-László, and Eric Bonabeau. 2003. "Scale-Free Networks." *Scientific American* 288 (5): 60–69.

Baran, Paul A., and Paul M. Sweezy. 1966. *Monopoly Capital: An Essay on the American Economic and Social Order*. New York: Monthly Review Press.

Barboza, David. 1998. "Farmers Are in Crisis as Hog Prices Collapse." *The New York Times*, December 13. http://nyti.ms/1D98Mz7.

Barker, Debbie. 2007. *The Rise and Predictable Fall of Globalized Industrial Agriculture*. San Francisco, CA: International Forum on Globalization.

Barker, Debbie, Bill Freese, and George Kimbrell. 2013. *Seed Giants vs. U.S. Farmers*. Washington, DC: Center for Food Safety.

Barlett, Donald L., and James B. Steele. 2008. "Monsanto's Harvest of Fear." *Vanity Fair*, May. http://www.vanityfair.com/politics/features/2008/05/monsanto200805.

Bartels, Larry M. 2010. *Unequal Democracy: The Political Economy of the New Gilded Age*. Princeton, NJ: Princeton University Press.

Bartz, Diane. 2014. "Sticker Shock Key to Antitrust Approval for Sysco, US Foods Deal." *Reuters*, January 31. http://trib.in/1q5CZ81.

Bartz, Diane. 2015. "Sysco Opposes FTC Challenge to Its Purchase of US Foods." *Reuters*, April 22. http://reut.rs/1HV0yxK.

BBC. 2012. "Farm Subsidy's Six-Figure Winners." *BBC News*, March 5. http://www.bbc.co.uk/news/uk-17225652.

Beer Insights. 2013. "Major Supplier Shipments and Share: 2012 vs 2011." http://www.beerinsights.com/.

Belasco, Warren J. 2007. *Appetite for Change: How the Counterculture Took On the Food Industry*. Ithaca, NY: Cornell University Press.

Bell, John. 2012. "Why Price Fixing Continues." *In the CEO Afterlife*. http://www.ceoafterlife.com/marketing/why-price-fixing-continues/.

Bellemare, Marc F. 2014. *Rising Food Prices, Food Price Volatility, and Social Unrest*. St. Paul, MN: Department of Applied Economics, University of Minnesota.

Belongia, Michael T. 1984. "The Dairy Price Support Program: A Study of Misdirected Economic Incentives." *Federal Reserve Bank of St. Louis Review*, February 5–14.

Belsie, Laurent. 2002. "Wal-Mart: World's Largest Company." *Christian Science Monitor*, February 19. http://www.csmonitor.com/2002/0219/p01s04-usec.html.

Berfield, Susan. 2013. "Where Wal-Mart Isn't: Four Countries the Retailer Can't Conquer." *Bloomberg BusinessWeek*, October 10. http://buswk.co/1tUYnV7.

Berlan, Jean-Pierre, and R. C. Lewontin. 1986. "The Political-Economy of Hybrid Corn." *Monthly Review* 38 (3): 35–47.

Bernhardt, Amy M., Cara Wilking, Anna M. Adachi-Mejia, Elaina Bergamini, Jill Marijnissen, and James D. Sargent. 2013. "How Television Fast Food Marketing Aimed at Children Compares with Adult Advertisements." *PLoS ONE* 8 (8): e72479.

Best, Henning. 2008. "Organic Agriculture and the Conventionalization Hypothesis: A Case Study from West Germany." *Agriculture and Human Values* 25 (1): 95–106.

Bichler, Shimshon, and Jonathan Nitzan. 2012. "The Asymptotes of Power." *Real-World Economics Review* 60: 18–53.

Bichler, Shimshon, and Jonathan Nitzan. 2014. "How Capitalists Learned to Stop Worrying and Love the Crisis." *Real-World Economics Review* 66: 65–73.

Black, Robert E., Cesar G. Victora, Susan P. Walker, Zulfiqar A. Bhutta, Parul Christian, Mercedes de Onis, Majid Ezzati, et al. 2013. "Maternal and Child Undernutrition and Overweight in Low-Income and Middle-Income Countries." *The Lancet* 382 (9890): 427–51.

Blackburn, Harvey, Carrie Welsh, and T. Stewart. 2005. "U.S. Swine Genetic Resources and the National Animal Germplasm Program." In *Swine Improvement Federation Proceedings*. Ottawa, ON, Canada. http://www.nsif.com/Conferences/2005/pdf%5CGermplasmProgram.pdf.

Blake, Andrew. 2003. "Syngenta Ties Seed Sales to Spray." *Farmers Weekly (UK)*, May 1. http://www.fwi.co.uk/article.asp?con=10148&sec=2&hier=66.

Blanding, Michael. 2011. "The Great Beer Challenge: How Two Companies Control Your Options." *Consumers Digest*, May. https://www.consumersdigest.com/special-reports/the-great-beer-challenge/view-all.

Blissett, Guy, Robin Kahn, and Maureen Stancik Boyce. 2008. *Break Out or Get Boxed In: Leading Strategies for Today's Food and Foodservice Distributors*. Somers, NY: IBM Corporation.

Blumenthal, Richard. 2014. "Blumenthal, Harkin Introduce Bill to End Federal Tax Subsidy for Unhealthy Food, Beverage Marketing to Children." http://1.usa.gov/1qXfseW.

Bonanno, Alessandro. 2009. "Sociology of Agriculture and Food Beginning and Maturity: The Contribution of the Missouri School." *Southern Rural Sociology* 24 (2): 29–47.

Bonanno, Alessandro, and Douglas H. Constance. 2010. *Stories of Globalization: Transnational Corporations, Resistance, and the State*. University Park, PA: Penn State Press.

Boshoff, Alison. 2013. "Supergreed." *Daily Mail*, June 14. http://dailym.ai/1gvARJ1.

Bousfield, Dan. 2013. "Fighting the Power? Struggle and Resistance in Capital as Power." In *The Capitalist Mode of Power: Critical Engagements with the Power Theory of Value*, edited by Tim Di Muzio, 103–16. New York: Routledge.

Bowles, Samuel, Richard Edwards, and Frank Roosevelt. 2005. *Understanding Capitalism: Competition, Command, and Change*. Oxford, UK: Oxford University Press.

Boyd, William. 2003. "Wonderful Potencies? Deep Structure and the Problem of Monopoly in Agricultural Biotechnology." In *Engineering Trouble: Biotechnology and Its Discontents*, edited by Rachel A. Schurman and Dennis Doyle Takahashi Kelso, 24–62. Berkeley, CA: University of California Press.

Boyle, Matthew. 2008. "Health Foods' Hidden Powerbroker." *CNN Money*, February 12. http://cnnmon.ie/1pKFEq1.

Bozic, Marin, John Newton, Andrew M. Novaković, Mark W. Stephenson, and Cameron S. Thraen. 2014. "Experts Analyze Dairy Policy in 2014 Farm Bill." *Farm and Dairy*, February 11. https://www.farmanddairy.com/news/experts-analyze-dairy-policy-2014-farm-bill/175478.html.

Braverman, Harry. 1998. *Labor and Monopoly Capital: The Degradation of Work in the Twentieth Century*. New York: NYU Press.

Brehm, Susan M. 2005. "From Red Barn to Facility: Changing Environmental Liability to Fit the Changing Structure of Livestock Production." *California Law Review* 93: 797–846.

Brewer, Annie Jean. 2011. "How to Make Homemade Soy Milk." *Yahoo! Voices*. http://voices.yahoo.com/how-homemade-soy-milk-7753079.html.

Brewers Association. 2014. "Craft Brewing Facts." http://www.brewersassociation.org/pages/business-tools/craft-brewing-statistics/facts.

Brock, James W. 2011. "Economic Concentration and Economic Power: John Flynn and a Quarter-Century of Mergers." *Antitrust Bulletin* 56 (4): 683–732.

Brown, Jason P., and Jeremy Glenn Weber. 2013. *The Off-Farm Occupations of U.S. Farm Operators and Their Spouses*. Economic Information Bulletin 156535. Washington, DC: USDA Economic Research Service. http://econpapers.repec.org/paper/agsu-ersib/156535.htm.

Brown, Lisa. 2014. "Is A-B InBev Brewing the Next Big Deal?" *St. Louis Post-Dispatch*, July 13. http://bit.ly/1nsFlvH.

Bruce, Amanda S., Jared M. Bruce, William R. Black, Rebecca J. Lepping, Janice M. Henry, Joseph Bradley, C. Cherry, Laura E. Martin, et al. 2014. "Branding and a Child's Brain: An fMRI Study of Neural Responses to Logos." *Social Cognitive and Affective Neuroscience* 9(1): 118–22.

Bucheli, Marcelo. 2008. "Multinational Corporations, Totalitarian Regimes and Economic Nationalism: United Fruit Company in Central America, 1899–1975." *Business History* 50 (4): 433–54.

Buck, Daniel, Christina Getz, and Julie Guthman. 1997. "From Farm to Table: The Organic Vegetable Commodity Chain of Northern California." *Sociologia Ruralis* 37 (1): 3–20.

Bugge, Annechen Bahr. 2011. "Lovin' It?: A Study of Youth and the Culture of Fast Food." *Food, Culture and Society: An International Journal of Multidisciplinary Research* 14 (1): 71–89.

Buono, Anthony F., and James L. Bowditch. 1989. *The Human Side of Mergers and Acquisitions: Managing Collisions Between People, Cultures, and Organizations*. San Francisco, CA: Jossey-Bass.

Burningham, Lucy. 2010. "Portland Food Carts Push Through Recession." *Oregon Business*, January. http://www.oregonbusiness.com/articles/78-january-2010/2775-cash-and-carry.

Burros, Marian. 1995. "U.S. Is Urged to Investigate Cereal Prices." *The New York Times*, March 8. http://nyti.ms/1rq9qVL.

Burros, Marian. 2008. "Supermarket Chains Narrow Their Sights." *The New York Times*, August 6. http://www.nytimes.com/2008/08/06/dining/06local.html.

Busch, Lawrence. 2011a. *Standards: Recipes for Reality*. Cambridge, MA: MIT Press.

Busch, Lawrence. 2011b. "Food Standards: The Cacophony of Governance." *Journal of Experimental Botany* 62 (10): 3247–50.

Cacace, L. Michael. 2014. "Fortune 500 2014 - Sysco." *Fortune*. http://fortune.com/fortune500/sysco-corporation-63/.

California v Safeway, Inc. 2011. US District Court, Central District of California.

Callicrate, Mike. 2015. "Merchants of Doubt Exposes the Bull." *No-Bull Food News*. http://nobull.mikecallicrate.com/2015/03/24/merchants-of-doubt-exposes-the-bull/.

Calvin, Linda, Roberta L. Cook, Mark Denbaly, Carolyn Dimitri, Lewrene Kay Glaser, Charles R. Handy, Mark D. Jekanowski, et al. 2001. *U.S. Fresh Fruit and Vegetable Marketing: Emerging Trade Practices, Trends, and Issues*. Agricultural Economics Reports 33915. United States Department of Agriculture, Economic Research Service. http://econpapers.repec.org/paper/agsuerser/33915.htm.

Campbell, Hugh, and Ruth Liepins. 2001. "Naming Organics: Understanding Organic Standards in New Zealand as a Discursive Field." *Sociologia Ruralis* 41 (1): 22–39.

Campbell, Hugh, and Christopher Rosin. 2011. "After the 'Organic Industrial Complex': An Ontological Expedition Through Commercial Organic Agriculture in New Zealand." *Journal of Rural Studies* 27 (4): 350–61.

Carlsson, Fredrik, Peter Frykblom, and Carl Johan Lagerkvist. 2007. "Consumer Willingness to Pay for Farm Animal Welfare: Mobile Abattoirs versus Transportation to Slaughter." *European Review of Agricultural Economics* 34(3): 321–44.

Carman, Tim. 2012. "Pork Industry Gives Sows Room to Move." *The Washington Post*, May 29. http://wapo.st/1m4Ij0N.

Carnevale, Chuck. 2014. "Sysco Is a Solid Company But Be Cautious When Buying Its Shares." *The Street*, June 27. http://bit.ly/1D99DA7.

Carney, Dan. 1995. "Dwayne's World." *Mother Jones*, July/August. http://www.mother-jones.com/politics/1995/07/dwaynes-world.

Carolan, Michael S. 2007. "Saving Seeds, Saving Culture: A Case Study of a Heritage Seed Bank." *Society & Natural Resources* 20 (8): 739–50.

Carolan, Michael. 2011. *The Real Cost of Cheap Food*. New York: Earthscan.

Carper, Jim. 2013. "Nestlé USA Takes the Top Spot on the Dairy 100." *Dairy Foods*, August 6.

Carson, Kevin A. 2007. *Studies in Mutualist Political Economy*. Charleston, SC: BookSurge Publishing.

Carson, Kevin A. 2008. *Organization Theory: A Libertarian Perspective*. Charleston, SC: BookSurge Publishing.

Carson Private Capital. 2009. "CPC Fund VI Announces Earthbound Farm Acquisition." http://www.carsoncapital.com/news07202009.shtml.

Carvajal, Doreen, and Stephen Castle. 2009. "A U.S. Hog Giant Transforms Eastern Europe." *The New York Times*, May 6. http://nyti.ms/1uyS20f.

Cavanaugh, Erica, Sarah Green, Giridhar Mallya, Ann Tierney, Colleen Brensinger, and Karen Glanz. 2014. "Changes in Food and Beverage Environments After an Urban Corner Store Intervention." *Preventive Medicine* 65: 7–12.

Chappell, Bill. 2013. "Home Brewing: Soon to Be Legal in All 50 States." *National Public Radio*, May 8. http://n.pr/1jyHAws.

Charterhouse Group. 2005. "Charterhouse Group Portfolio Company Creates a Leading Organic and Natural Baking Company." http://charterhousegroup.com/press_08-31-2005.html.

Chiquita. 2006. *Chiquita Brands International 2005 Annual Report*. Cincinnati, OH: Chiquita Brands International.

Choi, Candice. 2014. "Coke, Pepsi Dropping 'BVO' from All Drinks." *Associated Press*, May 5. http://yhoo.it/1tTOl2K.

Chug, Kiran. 2011. "Animal Death Toll Ends Cloning Trials." *Dominion Post*, February 21. http://bit.ly/1pbowMQ.

Chutchian, Maria. 2014. "Arbitration Loss Sends Swine Genetics Co. into Ch. 11." *Law360*, February 14. http://www.law360.com/articles/510414/arbitration-loss-sends-swine-genetics-co-into-ch-11.

Clapp, Jennifer. 2009. "Corporate Interests in US Food Aid Policy: Global Implications of Resistance to Reform." In *Corporate Power in Global Agrifood Governance*, edited by Jennifer Clapp and Doris A. Fuchs, 125–52. Cambridge, MA: MIT Press.

Clapp, Jennifer. 2014. "Financialization, Distance and Global Food Politics." *Journal of Peasant Studies* 41 (5): 797–814.

Clifford, Stephanie. 2011. "Wal-Mart Tests Online Grocery Service." *The New York Times*, April 24. http://nyti.ms/YK2Pt5.

Cloke, Paul. 1996. "Looking Through European Eyes? A Re-Evaluation of Agricultural Deregulation in New Zealand." *Sociologia Ruralis* 36 (3): 307–30.

Certified Naturally Grown. 2014. "Certified Naturally Grown: The Grassroots Alternative to Certified Organic." https://www.naturallygrown.org/.

Cochrane, D. T. 2010. "Review of Nitzan and Bichler's 'Capital as Power: A Study of Order and Creorder.'" *Theory in Action* 3 (2): 110–16.

Cochrane, Willard W. 1979. *The Development of American Agriculture: A Historical Analysis*. Minneapolis, MN: University of Minnesota Press.

Coffey, Brendan. 2013. "Sam Adams Creator Becomes Billionaire as Craft Beer Rises." *Bloomberg*, September 9. http://bloom.bg/1nAERul.

Coghlan, Andy, and Debora MacKenzie. 2011. "Revealed – the Capitalist Network That Runs the World." *New Scientist*, October 24. http://bit.ly/1nWnv6P.

Collings, Richard. 2013. "With Cash to Burn, Sysco Will Be Looking Out for Deals." *The Deal Pipeline*, February 22. http://bit.ly/1m79FD9.

Collins, Laura. 2014. "The 2014 Farm Bill Subsidy Reforms Don't Go Far Enough." *American Action Forum*. http://bit.ly/1dcJ7tD.

Compa, Lance. 2004. *Blood, Sweat, and Fear: Workers' Rights in U.S. Meat and Poultry Plants*. New York: Human Rights Watch.

CompanyPay.com. 2014. "Executive Compensation." http://www.companypay.com/exec-utive/compensation/retail_grocery_stores.asp.

Condra, Alli. 2013. *Cottage Food Laws in the United States*. Cambridge, MA: Harvard Food Law and Policy Clinic.

Connor, John, Peter C. Carstensen, Roger A. McEowen, and Neil E. Harl. 2002. *The Ban on Packer Ownership and Feeding of Livestock: Legal and Economic Implications*. Ames, IA: Department of Economics, Iowa State University. http://www.econ.iastate.edu/sites/default/files/publications/papers/p7173-2002-03-01.pdf.

Connor, John M. 2007. *Global Price Fixing*. 2nd ed. New York: Springer.

Constance, Douglas H., and Alessandro Bonanno. 1999. "CAFO Controversy in the Texas Panhandle Region: The Environmental Crisis of Hog Production." *Culture & Agriculture* 21 (1): 14–26.

Constance, Douglas H., Jin Young Choi, and Holly Lyke-Ho-Gland. 2008. "Conventionalization, Bifurcation, and Quality of Life: Certified and Non-Certified Organic Farmers in Texas." *Southern Rural Sociology* 23 (1): 208–34.

Constance, Douglas H., Mary Hendrickson, and Philip H. Howard. 2014. "Agribusiness Concentration: Globalization, Market Power, and Resistance." In *The Global Food System: Issues and Solutions*, edited by Schanbacher, William D., 31–58. Santa Barbara, CA: ABC-CLIO.

Constance, Douglas H., Mary Hendrickson, Philip H. Howard, and William D. Heffernan. 2014. "Economic Concentration in the Agrifood System: Impacts on Rural Communities and Emerging Responses." In *Rural America in a Changing World: Problems and Prospects for the 2010s*, edited by Bailey, Conner, Jensen, Leif, and Ransom, Elizabeth, 18–35. Morgantown, WV: West Virginia University Press.

Cook, Christopher D. 2015. "Seed Libraries Fight for the Right to Share." *Shareable*, February 11. http://www.shareable.net/blog/seed-libraries-fight-for-the-right-to-share

Cook, Roberta L. 2012. *Trends in the Marketing of Fresh Produce and Fresh-Cut/Value-Added Produce*. Davis, CA: University of California, Davis.

Cornucopia Institute. 2013. *FDA/USDA Collude to Eliminate True Organic Egg Production*. Cornucopia, WI: Cornucopia Institute. http://bit.ly/1y4oRGs

Cornucopia Institute. 2014. "Leading Organic Brand, Horizon, Blasted for Betraying Organics." *Cornucopia Institute*. http://bit.ly/X4lYnD/.

Corporate Crime Reporter. 2010. "Antitrust Probe of Monsanto May Turn on Anti-Stacking Provisions." http://www.corporatecrimereporter.com/monsanto051010.htm.

Cotterill, Ronald W. 1999. "Jawboning Cereal: The Campaign to Lower Cereal Prices." *Agribusiness* 15 (2): 197–205.

Covington, Calvin. 2013. "Dairy Breeds Have Shifted with the Milk Markets." *Hoard's Dairyman*, November.

Cox, Craig. 2013. "Programs to Reduce Ag's Water Use Must Be Strengthened, Not Cut." *Environmental Working Group*. http://bit.ly/1tV0zfg.

Crawley, Nancy. 2009. "At 75, Meijer Family Still Controls Retailer's Destiny." *Grand Rapids Press*, June 28. http://bit.ly/1I4jmMr.

Cray, Charlie. 2010. "Banana Land and the Corporate Death Squad Scandals." *CorpWatch*, February 25. http://www.corpwatch.org/article.php?id=15535.

Crotty, Ann. 2013. "African Tax Treaty to Combat Evasion." *Independent Online*, July 8. http://bit.ly/YK3pHi.

Crowell, Chris. 2013. "Craft Beer Distribution: Study the Market, Distributors and Your Own Operations." *Craft Brewing Business*, September 10. http://bit.ly/1lxdZ7h.

CSP Daily News. 2013. "What the Shelves Can Teach Us." October 21. http://bit.ly/1HZK5lk.

Culliney, Kacey. 2013. "Canada Price Fixing Woes: Chocolate Titans Settle Class Action but Still Face Criminal Charges." *Confectionery News*, September 19. http://bit.ly/1sUg9pM.

Cummins, Ronnie. 2006. "USDA Attempts to Pack Organic Standards Board with Corporate Agribusiness Reps: Organic Consumers Fight Hijacked Seats on NOSB." *Organic Consumers Association*. http://www.organicconsumers.org/articles/article_3526.cfm.

Cummins, Steven, Ellen Flint, and Stephen A. Matthews. 2014. "New Neighborhood Grocery Store Increased Awareness of Food Access But Did Not Alter Dietary Habits Or Obesity." *Health Affairs* 33 (2): 283–91.

Cummins, Steven, and Sally Macintyre. 2002. "'Food Deserts'—Evidence and Assumption in Health Policy Making." *BMJ* 325 (7361): 436–38.

Cummins, Steven, and Sally Macintyre. 2006. "Food Environments and Obesity—Neighbourhood or Nation?" *International Journal of Epidemiology* 35 (1): 100–104.

Dalton, Jen. 2011. "Faces & Visions of the Food Movement: Rick North." *Civil Eats*. http://civileats.com/2011/01/03/faces-visions-of-the-food-movement-rick-north/.

Davis, Brennan, and Christopher Carpenter. 2009. "Proximity of Fast-Food Restaurants to Schools and Adolescent Obesity." *American Journal of Public Health* 99 (3): 505–10.

Day, Nicholas. 2006. "Bye-Bye Bell's." *Chicago Reader*. http://www.chicagoreader.com/chicago/bye-bye-bells/Content?oid=923858.

Dayen, David. 2014. "The Farm Bill Still Gives Wads of Cash to Agribusiness. It's Just Sneakier About It." *The New Republic*, February 7. http://bit.ly/1kDRU9O.

De Ferranti, Sarah D. 2012. "Declining Cholesterol Levels in US Youths: A Reason for Optimism." *JAMA* 308 (6): 621–22.

De Schutter, Olivier. 2014. *The Transformative Potential of the Right to Food*. New York: United Nations Human Rights Council.

Deierlein, Andrea L., Maida P. Galvez, Irene H. Yen, Susan M. Pinney, Frank M. Biro, Lawrence H. Kushi, Susan Teitelbaum, and Mary S. Wolff. 2014. "Local Food Environments Are Associated with Girls' Energy, Sugar-Sweetened Beverage and Snack-Food Intakes." *Public Health Nutrition* 17 (10): 2194–200.

Deighton, John. 1999. *Snapple*. 9-599-126. Cambridge, MA: Harvard Business School Case, Harvard University.

DeLind, Laura B. 1995. "The State, Hog Hotels, and the 'Right to Farm': A Curious Relationship." *Agriculture and Human Values* 12 (3): 34–44.

DeLind, Laura B. 2011. "Are Local Food and the Local Food Movement Taking Us Where We Want to Go? Or Are We Hitching Our Wagons to the Wrong Stars?" *Agriculture and Human Values* 28 (2): 273–83.

DeLind, Laura B., and Philip H. Howard. 2008. "Safe at Any Scale? Food Scares, Food Regulation, and Scaled Alternatives." *Agriculture and Human Values* 25 (3): 301–17.

Department of Justice. 2011a. "Justice Department Reaches Settlement with Dean Foods Company." http://www.justice.gov/opa/pr/2011/March/11-at-388.html.

Department of Justice. 2011b. "Statement on the Department of Justice's Antitrust Division on Its Decision to Close Its Investigation of Perdue's Acquisition of Coleman Natural Foods." http://www.justice.gov/atr/public/press_releases/2011/270591.htm.

Department of Justice. 2012. *Competition and Agriculture: Voices from the Workshops on Agriculture and Antitrust Enforcement in Our 21st Century Economy and Thoughts on the Way Forward*. Washington, DC: Department of Justice.

DeVore, Brian. 2012. "Crop Insurance: A Safety Net Becomes a Threat." *Land Stewardship Project*. http://landstewardshipproject.org/posts/281.

Dezember, Ryan. 2012. "The Buyout Brain Behind Annie's IPO." *Wall Street Journal*, April 13. http://on.wsj.com/1ok4Hzc.

Di Muzio, Tim. 2013. "The Provocations of Capital as Power." In *The Capitalist Mode of Power: Critical Engagements with the Power Theory of Value*, edited by Tim Di Muzio, 1–16. New York: Routledge.

Dickerson, Marla 2004. "Small Tortilla Makers Lose Antitrust Suit Against Rival." Los Angeles Times, January 6. http://lat.ms/1BnJ10n

Dillon, Matthew. 2005. "Monsanto Buys Seminis." *New Farm*, February 22. http://newfarm.rodaleinstitute.org/features/2005/0205/seminisbuy/.

Dillon, Matthew. 2008. "Another Big Horticultural Seed Company Bought by Monsanto." *Grist*. http://grist.org/article/who-owns-your-tomato/.

Dillon, Matthew. 2010. "Organic Vegetable Farmers – WARNING – You May Be Engaging in Contract Agreements with Monsanto." *Seed Broadcast*. http://bit.ly/1FEdbgd.

DiLorenzo, Thomas J. 1996. "The Myth of Natural Monopoly." *The Review of Austrian Economics* 9 (2): 43–58.

Dixon, Jane. 2002. *The Changing Chicken: Chooks, Cooks and Culinary Culture*. Sydney, Australia: UNSW Press.

Dobrow, Joe. 2014. *Natural Prophets: From Health Foods to Whole Foods*. New York: Rodale.

Dobson, Paul W., Michael Waterson, and Stephen W. Davies. 2003. "The Patterns and Implications of Increasing Concentration in European Food Retailing." *Journal of Agricultural Economics* 54 (1): 111–25.

DOJ USDA. 2010. *Public Workshops Exploring Competition Issues in Agriculture: A Dialogue on Competition Issues Facing Farmers in Today's Agricultural Marketplace*. Ankeny, IA: US Department of Justice, US Department of Agriculture.

Dolan, Kerry A., and Luisa Kroll. 2014. "The Richest People on the Planet 2014." *Forbes*, March 3. http://www.forbes.com/billionaires/.

Domhoff, G. William. 2014. *Who Rules America? The Triumph of the Corporate Rich*. 7th ed. New York: McGraw-Hill Higher Education.

Domina, David A. 2004. "Proving Anti-Competitive Conduct in the U.S. Courtroom: The Plaintiff's Argument in Pickett v Tyson Fresh Meats, Inc." *Journal of Agricultural & Food Industrial Organization* 2 (1): Article 8, 1–32.

Domina, David A., and C. Robert Taylor. 2009. "The Debilitating Effects of Concentration Markets Affecting Agriculture." *Drake Journal of Agricultural Law* 15: 61–108.

Dooley, Roger. 2011. *Brainfluence: 100 Ways to Persuade and Convince Consumers with Neuromarketing*. New York: John Wiley & Sons.

Doyle, Michael. 2013. "USDA Hopes to Settle Discrimination Suits by Hispanic and Female Farmers." *McClatchy Newspapers*, April 9. http://bit.ly/1uyT3p2.

Du Boff, Richard B., and Edward S. Herman. 2001. "Mergers, Concentration, and the Erosion of Democracy." *Monthly Review* 53 (1): 14–29.

Duggan, Tara. 2014. "New 'Girl' is a Monsanto-Free Tomato." *San Francisco Chronicle*, March 7. http://bit.ly/1qCmiqU.

DuPont. 2008. "General Mills and DuPont Sell 8th Continent Joint Venture to Stremicks Heritage Foods." http://prn.to/1s2lj3l.

Dupraz, Emily. 2012. *Monsanto and the Per Se Illegal Rule for Bundled Discounts*. South Royalton, VT: Vermont Law School. http://papers.ssrn.com/abstract=2032615.

DuPuis, E. Melanie, and Sean Gillon. 2009. "Alternative Modes of Governance: Organic as Civic Engagement." *Agriculture and Human Values* 26 (1): 43–56.

Dyer, Jessica. 2010. "Organic Seed Firm to Relocate." *Albuquerque Journal*, August 14. http://www.abqjournal.com/north/1401732north08-14-10.htm.

Easley, David, and Jon Kleinberg. 2010. *Networks, Crowds, and Markets: Reasoning About a Highly Connected World*. New York: Cambridge University Press.

Eden Foods. 2005. "Communiqué on OTA's Consistent Work to Weaken Organic Standards." http://www.edenfoods.com/articles/view.php?articles_id=70.

Eden Foods. 2006. "Why Eden Foods Chooses Not to Use the USDA Seal." http://www.edenfoods.com/articles/view.php?articles_id=78.

Eisenhauer, Elizabeth. 2001. "In Poor Health: Supermarket Redlining and Urban Nutrition." *GeoJournal* 53 (2): 125–33.

Elinder, Liselotte Schäfer. 2005. "Obesity, Hunger, and Agriculture: The Damaging Role of Subsidies." *British Medical Journal* 331 (7528): 1333–36.

Environmental Working Group. 2013. "2013 Farm Subsidy Database." http://farm.ewg.org/.

Equation Research. 2005. *Organic Trend Tracker*. http://www.webwire.com/ViewPressRel.asp?aId=5889.

Erickson, Gary. 2004. *Raising the Bar: Integrity and Passion in Life and Business: The Story of Clif Bar Inc*. New York: Wiley.

Estabrook, Barry. 2010. "A Tale of Two Dairy Farms." *The Atlantic*, August 10. http://theatln.tc/1qVyFMx.

Estabrook, Barry. 2013. "Sustainable Pork Farming Is Real." *OnEarth*, April 14. http://www.onearth.org/article/meet-the-farmer-selling-chipotle-antibiotic-free-pork.

ETC Group. 2013a. *Gene Giants Seek "Philanthrogopoly."* Communique No. 110. Ottawa, ON, Canada.

ETC Group. 2013b. *Putting the Cartel Before the Horse … and Farm, Seeds, Soil, Peasants, Etc*. Communique No. 111. Ottawa, ON, Canada.

Etter, Lauren. 2010. "Plenty of Spilled Milk to Cry Over for Dairymen Lured to U.S." *Wall Street Journal*, February 16. http://on.wsj.com/1wlnsH0.

Etter, Lauren, and John Lyons. 2008. "Brazilian Beef Clan Goes Global as Troubles Hit Market." *Wall Street Journal*, August 1. http://on.wsj.com/1m7aDiJ.

Faillace, Linda. 2006. *Mad Sheep: The True Story Behind the USDA's War on a Family Farm*. White River Junction, VT: Chelsea Green Publishing.

Falat, Stacia Marie. 2011. *Scaling up 'Buy Local, Sell Fresh.'* East Lansing, MI: Michigan State University.

Fantle, Will. 2014. "Legal Complaint Against Horizon," February 11. http://www.cornucopia.org/USDA/HorizonUSDA_LegalComplaint2014.pdf.

Farah, George, and Justin Elga. 2001. "What's Not Talked About on Sunday Morning?" *Extra!*, October. http://fair.org/extra-online-articles/whats-inoti-talked-about-on-sunday-morning/.

Farm and Dairy. 2002. "Court Rules Against Pork Checkoffs." *Farm and Dairy*, October 31. http://www.farmanddairy.com/news/court-rules-against-pork-checkoffs/1198.html.

Farrell, Joseph, and Carl Shapiro. 2001. "Scale Economies and Synergies in Horizontal Merger Analysis." *Antitrust Law Journal* 68 (3): 685–710.

Fatka, Jacqui. 2007. "Dow Increasing Seed Market Share." *Feedstuffs*, September 17.

Federal Trade Commission. 2001. *Report on the Federal Trade Commission Workshop on Slotting Allowances and Other Marketing Practices in the Grocery Industry*. Washington, DC: Federal Trade Commission.

Feeding America. 2014. *Hunger in America 2014*. Chicago, IL: Feeding America.

Fernandez-Cornejo, Jorge. 2004. *The Seed Industry in U.S. Agriculture: An Exploration of Data and Information on Crop Seed Markets, Regulation, Industry Structure, and Research and Development*. Washington, DC: Agricultural Information Bulletin 786. USDA Economic Research Service.

Fernandez-Cornejo, Jorge, and Richard E. Just. 2007. "Researchability of Modern Agricultural Input Markets and Growing Concentration." *American Journal of Agricultural Economics* 89 (5): 1269–75.

Fetter, T. Robert, and Julie A. Caswell. 2002. "Variation in Organic Standards Prior to the National Organic Program." *American Journal of Alternative Agriculture* 17 (2): 55–74.

Financial Times. "Global 500." June 27, 2014. http://on.ft.com/1FCqpFK.

Fishman, Charles. 2003. "The Wal-Mart You Don't Know." *Fast Company*, December 1. http://www.fastcompany.com/47593/wal-mart-you-dont-know.

Fleischhacker, S. E., K. R. Evenson, D. A. Rodriguez, and A. S. Ammerman. 2011. "A Systematic Review of Fast Food Access Studies." *Obesity Reviews* 12 (5): e460–e471.

Food & Water Watch. 2008. *The Trouble with Smithfield: A Corporate Profile*. Washington, DC: Food & Water Watch. http://documents.foodandwaterwatch.org/doc/SmithfieldJan08.pdf.

Food & Water Watch. 2013. *Grocery Goliaths: How Food Monopolies Impact Consumers*. Washington, DC: Food & Water Watch. http://documents.foodandwaterwatch.org/doc/grocery_goliaths.pdf.

Forbes. 2014. "Performance Food Group." *Forbes*. http://www.forbes.com/companies/performance-food-group/.

Ford, Paula B., and David A. Dzewaltowski. 2008. "Disparities in Obesity Prevalence Due to Variation in the Retail Food Environment: Three Testable Hypotheses." *Nutrition Reviews* 66(4): 216–28.

Foster, John Bellamy, and Robert W. W. McChesney. 2012. *The Endless Crisis: How Monopoly-Finance Capital Produces Stagnation and Upheaval from the USA to China*. New York: NYU Press.

Fowler, Cary, and Patrick R. Mooney. 1990. *Shattering: Food, Politics, and the Loss of Genetic Diversity*. Tucson, AZ: University of Arizona Press.

Franck, Caroline, Sonia M. Grandi, and Mark J. Eisenberg. 2013. "Agricultural Subsidies and the American Obesity Epidemic." *American Journal of Preventive Medicine* 45 (3): 327–33.

Franco, Michael. 2014. "Cap'n Crunch is Staring at Your Kid for a Reason." *CNET*, April 2. http://cnet.co/1kqjR1Q.

Frankel, Todd C. 2010. "A-B to Increase Prices Despite Soft U.S. Sales." *STLtoday.com*, August 13. http://bit.ly/TNy5Vs.

Fraser, Rebekah L. 2009. "Seed Research: Seed Cleaning Takes a Bath." *Growing Magazine*, July. http://www.growingmagazine.com/article-3671.aspx.

Freese, Betsy. 2013. "Pork Powerhouses 2013." *Successful Farming*, September 30. http://bit.ly/1Pda2Vd.

Freudenburg, William R., and Margarita Alario. 2007. "Weapons of Mass Distraction: Magicianship, Misdirection and the Dark Side of Legitimation." *Sociological Forum* 22 (2): 146–173.

Fridell, Gavin, and Martijn Konings. 2013. *Age of Icons: Exploring Philanthrocapitalism in the Contemporary World*. Toronto, ON, Canada: University of Toronto Press.

Friedland, William H. 1984. "Commodity Systems Analysis: An Approach to the Sociology of Agriculture." In *Research in Rural Sociology and Development*, edited by Schwarzweller, Harry K., 221–35. London: JAI Press.

Friedland, William H. 2004. "Agrifood Globalization and Commodity Systems." *International Journal of Sociology of Agriculture and Food* 12 (1): 5–16.

Friedland, William H., Amy E. Barton, and Robert J. Thomas. 1981. *Manufacturing Green Gold: Capital, Labor, and Technology in the Lettuce Industry*. Cambridge, UK: Cambridge University Press.

Fromartz, Samuel. 2006. *Organic, Inc.: Natural Foods and How They Grew*. Orlando, FL: Harcourt.

Fryar, Cheryl D., and R. Bethene Ervin. 2013. *Caloric Intake from Fast Food Among Adults: United States, 2007–2010*. Data Brief 114. Hyattsville, MD: National Center for Health Statistics.

Fuglie, Keith, Paul Heisey, John L. King, Kelly Day-Rubenstein, David Schimmelpfennig, Sun Ling Wang, Carl E. Pray, and Rupa Karmarkar-Deshmukh. 2011. *Research Investments and Market Structure in the Food Processing, Agricultural Input, and Biofuel Industries Worldwide*. Economic Research Report - 130. Washington, DC: USDA Economic Research Service.

Fulmer, Melinda. 2000. "Cheese Makers Face Antitrust Allegations." *Los Angeles Times*, December 2. http://articles.latimes.com/2000/dec/02/business/fi-60204.

Furnari, Chris. 2014. "Brewers, Distributors Discuss Franchise Law Reform at NBWA Legislative Conference." *Brewbound*, April 28. http://bit.ly/1bnuVzP.

Gadiesh, Orit, Charles Ormiston, Sam Rovit, and Julian Critchlow. 2001. "The 'why' and 'how' of Merger Success." *European Business Journal* 13 (4): 187–93.

Galbraith-Emami, Sarah, and Tim Lobstein. 2013. "The Impact of Initiatives to Limit the Advertising of Food and Beverage Products to Children: A Systematic Review." *Obesity Reviews* 14 (12): 960–74.

Gallant, Andre. 2013. "Why Lettuce Keeps Making Us Sick." *Modern Farmer*, July 16. http://modernfarmer.com/2013/07/why-lettuce-keeps-making-us-sick/.

Galt, Ryan E. 2013. "Placing Food Systems in First World Political Ecology: A Review and Research Agenda." *Geography Compass* 7 (9): 637–58.

Garrison-Sprenger, Nicole. 2003. "General Mills Sweet on Soymilk." *Minneapolis/St. Paul Business Journal*, November 23. http://bit.ly/1m7b2Si.

Gasparro, Annie, and Leslie Josephs. 2015. "Whole Foods Calls the Shots for Startups." *The Wall Street Journal*, May 6. http://on.wsj.com/1EQtVib.

Geiger, Lynn. 2014. "Beyond Local." *Traverse City Business News*, February. http://www.tcbusinessnews.com/news/beyond-local.

George, Lisa M. 2011. "The Growth of Television and the Decline of Local Beer." In *The Economics of Beer*, edited by Johan F. M. Swinnen, 213–26. Oxford, UK: Oxford University Press.

George, Susan. 1999. "A Short History of Neo-Liberalism: Twenty Years of Elite Economics and Emerging Opportunities for Structural Change." Presented at the Conference on Economic Sovereignty in a Globalising World, Bangkok, Thailand, March.

Gereffi, Gary, and Michelle Christian. 2009. "The Impacts of Wal-Mart: The Rise and Consequences of the World's Dominant Retailer." *Annual Review of Sociology* 35 (1): 573–91.

Gereffi, Gary, Joonkoo Lee, and Michelle Christian. 2009. "US-Based Food and Agricultural Value Chains and Their Relevance to Healthy Diets." *Journal of Hunger & Environmental Nutrition* 4 (3–4): 357–74.

Gerlock, Grant. 2014. "Hog Farmers Differ on Packer-Owned Pigs." *Iowa Public Radio*, March 26. http://iowapublicradio.org/post/hog-farmers-differ-packer-owned-pigs.

Gibson-Graham, J. K. 2006. *The End of Capitalism (As We Knew It): A Feminist Critique of Political Economy*. Minneapolis, MN: University of Minnesota Press.

Gilens, Martin. 2014. *Affluence and Influence: Economic Inequality and Political Power in America*. Princeton, NJ: Princeton University Press.

Gillam, Carey. 2013. "Monsanto, DuPont Strike $1.75 Billion Licensing Deal, End Lawsuits." *Reuters*, March 26. http://reut.rs/1sUj9CF.

Gillam, Carey. 2015. "Kraft Removing Synthetic Colors from Iconic Macaroni & Cheese." *Reuters*, April 20. http://reut.rs/1EkRx1h.

Giroux, Greg. 2013. "Grocers' Group Spends Record Lobbying Amid Food-Labeling Fights." *Business Week*, December 3. http://buswk.co/1q5H1gB.

Glaser, Lewrene K., and Gary D. Thompson. 2001. *The U.S. Lettuce and Fresh-Cut Vegetable Industries: Marketing Channels, Sales Arrangements, Fees, and Services*. VGS-283. Washington, DC: USDA Economic Research Service.

Goeden, Gerry. 2014. "What Will the World Inherit from GE Salmon." *Independent Science News*, May 12. http://bit.ly/1jlFoaJ.

Goldman, Seth. 2010. "Changing the Status Quo." *Inc.*, April 28. http://www.inc.com/seth-goldman/changing-the-status-quo.html.

Goldschmidt, Walter Rochs. 1978. *As You Sow: Three Studies in the Social Consequences of Agribusiness*. New York: Harcourt.

Goodman, David. 2002. "Rethinking Food Production–Consumption: Integrative Perspectives." *Sociologia Ruralis* 42 (4): 271–77.

Goodman, David. 2003. "The Quality 'Turn' and Alternative Food Practices: Reflections and Agenda." *Journal of Rural Studies* 19 (1): 1–7.

Goodman, David, and E. Melanie DuPuis. 2002. "Knowing Food and Growing Food: Beyond the Production–Consumption Debate in the Sociology of Agriculture." *Sociologia Ruralis* 42 (1): 5–22.

Goodman, David, E., Melanie DuPuis, and Michael K. Goodman. 2012. *Alternative Food Networks: Knowledge, Practice, and Politics*. New York: Routledge.

Goodman, David, Bernardo Sorj, and John Wilkinson. 1987. *From Farming to Biotechnology: A Theory of Agro-Industrial Development*. Oxford: Basil Blackwell.

Goodwin, Barry K., Ashok K. Mishra, and François Ortalo-Magné. 2011. *The Buck Stops Where? The Distribution of Agricultural Subsidies*. Working Paper 16693. Cambridge, MA: National Bureau of Economic Research. http://www.nber.org/papers/w16693.

Goodwyn, Lawrence. 1978. *The Populist Moment: A Short History of the Agrarian Revolt in America: A Short History of the Agrarian Revolt in America*. Oxford: Oxford University Press.

Gould, Brian W. 2010. "Consolidation and Concentration in the U.S. Dairy Industry." *Choices* 25 (2): 1–15.

Goulson, Dave. 2013. "An Overview of the Environmental Risks Posed by Neonicotinoid Insecticides." *Journal of Applied Ecology* 50 (4): 977–87.

GRAIN. 2012. *The Great Food Robbery: How Corporations Control Food, Grab Land and Destroy the Climate*. Barcelona, Spain: Genetic Resources Action International.

Gramsci, Antonio. 1971. *Selections from the Prison Notebooks of Antonio Gramsci*. New York: International Publishers.

Gray, Thomas W., William D. Heffernan, and Mary K. Hendrickson. 2001. "Agricultural Cooperatives and Dilemmas of Survival." *Journal of Rural Cooperation* 29 (2): 167–92.

Gray, W. Blake. 2013. "Is a Budweiser/Corona Merger Really Bad for America?" *The Gray Report*. http://blog.wblakegray.com/2013/02/is-budweisercorona-merger-really-bad.html.

Green, Gary P., and William D. Heffernan. 1984. "Economic Dualism in American Agriculture." *Southern Rural Sociology* 2 (1): 201–10.

Greenhouse, Steven. 2014. "In Florida Tomato Fields, a Penny Buys Progress." *The New York Times*, April 24. http://nyti.ms/1nMTB5Q.

Greenwald, Glenn. 2014. "The U.S. Government's Secret Plans to Spy for American Corporations." *The Intercept*, September 5. http://bit.ly/1ILLs5s.

Grey, Mark. 2000. "'Those Bastards Can Go to Hell!' Small-Farmer Resistance to Vertical Integration and Concentration in the Pork Industry." *Human Organization* 59 (2): 169–76.

Griffioen, James. 2011. "Yes There Are Grocery Stores in Detroit." *The Urbanophile*. http://www.urbanophile.com/2011/01/25/yes-there-are-grocery-stores-in-detroit-by-james-griffioen/.

Grimaud. 2014. "Groupe Grimaud." http://www.grimaud.com/en/.

Grimes, Warren S. 1999. "Market Definition in Franchise Antitrust Claims: Relational Market Power and the Franchisor's Conflict of Interest." *Antitrust Law Journal* 67 (2): 243–81.

Grocery Headquarters. 2013. "Top Margarine/Spread/Butter Blend Makers, 2012." from SymphonyIRI Group Inc.

Grossberg, Josh. 2002. "Minority Report's Product Placement." *E! Entertainment*, June 21. http://www.eonline.com/news/43478/minority-reports-product-placement.

Guebert, Alan. 2012. "Farm & Food: The Dairy Deals." *Peoria Journal Star*, November 3. http://bit.ly/1qVzB3A.

Gumpert, David E. 2009. *The Raw Milk Revolution: Behind America's Emerging Battle Over Food Rights*. White River Junction, VT: Chelsea Green Publishing.

Gumpert, David. 2014. "MI Judge Ruling: Mark Baker Can Raise His 'Feral' Pigs, But What About Everyone Else?" *The Complete Patient*. http://bit.ly/1sUjXao.

Guptill, Amy. 2009. "Exploring the Conventionalization of Organic Dairy: Trends and Counter-Trends in Upstate New York." *Agriculture and Human Values* 26 (1–2): 29–42.

Guptill, Amy, and Rick Welsh. 2014. "The Declining Middle of American Agriculture: A Spatial Phenomenon." In *Rural America in a Changing World: Problems and Prospects for the 2010s*, edited by Bailey, Conner, Jensen, Leif, and Ransom, Elizabeth, 36–50. Morgantown, WV: West Virginia University Press.

Gura, Susanne. 2007. *Livestock Genetics Companies: Concentration and Proprietary Strategies of an Emerging Power in the Global Food Economy*. Ober-Ramstadt, Germany: League for Pastoral Peoples and Endogenous Livestock Development.

Gurian-Sherman, Doug. 2008. *CAFOs Uncovered: The Untold Costs of Confined Animal Feeding Operations*. Cambridge, MA: Union of Concerned Scientists.

Gustafson, Katherine. 2012. *Change Comes to Dinner: How Vertical Farmers, Urban Growers, and Other Innovators Are Revolutionizing How America Eats*. New York: St. Martin's Press.

Guthman, Julie. 1998. "Regulating Meaning, Appropriating Nature: The Codification of California Organic Agriculture." *Antipode* 30 (2): 135–54.

Guthman, Julie. 2003. "Fast Food/Organic Food: Reflexive Tastes and the Making of Yuppie Chow." *Social & Cultural Geography* 4 (1): 45–58.

Guthman, Julie. 2004a. *Agrarian Dreams: The Paradox of Organic Farming in California*. Berkeley, CA: University of California Press.

Guthman, Julie. 2004b. "The Trouble with 'Organic Lite' in California: A Rejoinder to the 'Conventionalisation' Debate." *Sociologia Ruralis* 44 (3): 301–16.

Gutknecht, Dave. 2003. "Co-Op Devolution Part 2: Northeast Cooperatives to Fold, United Natural Foods, Inc. Assuming Services After Merger Vote." *Cooperative Grocer*. http://www.cooperativegrocer.coop/articles/index.php?id=424.

Hackbarth, Diana P., Barbara Silvestri, and William Cosper. 1995. "Tobacco and Alcohol Billboards in 50 Chicago Neighborhoods: Market Segmentation to Sell Dangerous Products to the Poor." *Journal of Public Health Policy* 16 (2): 213–30.

Haderspeck, Jennifer. 2012. "Nothing Alternative about Dairy Alternatives." *Beverage Industry*, November. http://bit.ly/1DIQh11.

Hager, Sandy Brian. 2013. "The Power of Investment Banks: Surplus Absorption or Differential Capitalization?" In *The Capitalist Mode of Power: Critical Engagements with the Power Theory of Value*, edited by Tim Di Muzio, 39–58. New York: Routledge.

Hall, Alan, and Veronika Mogyorody. 2001. "Organic Farmers in Ontario: An Examination of the Conventionalization Argument." *Sociologia Ruralis* 41 (4): 399–22.

Hamilton, Lisa M. 2014. "Linux for Lettuce." *VQR: A National Journal of Literature & Discussion*, May 14. http://www.vqronline.org/reporting-articles/2014/05/linux-lettuce.

Hannaford, Steve. 2007. *Market Domination! The Impact of Industry Consolidation on Competition, Innovation, and Consumer Choice*. Westport, CT: Greenwood Publishing Group.

Hardin, Pete. 2010. "How Did Dairy End Up in This Crooked Mess?" *The Milkweed*, June.

Harl, Neil E. 2000. "The Age of Contract Agriculture: Consequences of Concentration in Input Supply." *Journal of Agribusiness* 18 (1): 115–27.

Harper, Will. 2005. "The O Word." *East Bay Express*, January 5. http://www.eastbayexpress.com/eastbay/the-o-word/Content?oid=1076351.

Hartnett, Kevin. 2014. "'Seed Libraries' Try to Save the World's Plants." *The Boston Globe*, March 9. http://bit.ly/1BEVZD4.

Harvey, Fiona. 2012. "'Just-in-Time' Business Models Put UK at Greater Risk in Event of Disasters, Warns Think Tank." *The Guardian*, January 5. http://bit.ly/1m4Ka5C.

Hasler, John F. 2003. "The Current Status and Future of Commercial Embryo Transfer in Cattle." *Animal Reproduction Science* 79 (3): 245–64.

Haumann, Barbara. 2014. *American Appetite for Organic Products Breaks Through $35 Billion Mark*. Brattleboro, VT: Organic Trade Association. http://bit.ly/1rCwaMC.

Hauter, Wenonah. 2012. *Foodopoly: The Battle Over the Future of Food and Farming in America*. New York: The New Press.

Hauter, Wenonah. 2014. "Food and Water Watch Slams Sysco-US Foods Merger," January 8. http://bit.ly/1c4t9Vy.

He, Meizi, Patricia Tucker, Jennifer D. Irwin, Jason Gilliland, Kristian Larsen, and Paul Hess. 2012. "Obesogenic Neighbourhoods: The Impact of Neighbourhood Restaurants and Convenience Stores on Adolescents' Food Consumption Behaviours." *Public Health Nutrition* 15 (12): 2331–39.

Hearson, Martin, and Richard Brooks. 2012. *Calling Time: Why SABMiller Should Stop Dodging Taxes in Africa*. London: ActionAid UK.

Heffernan, William D. 1972. "Sociological Dimensions of Agricultural Structures in the United States." *Sociologia Ruralis* 12 (2): 481–99.

Heffernan, William D. 1999. "Biotechnology and Mature Capitalism." Presented at the 11th Annual Meeting of the National Agricultural Biotechnology Council, Lincoln, NE, June.

Heffernan, William D. 2000. "Concentration of Ownership and Control in Agriculture." In *Hungry for Profit: The Agribusiness Threat to Farmers, Food, and the Environment*, edited by Fred Magdoff, John Bellamy Foster, and Frederick H. Buttel, 61–75. New York: Monthly Review Press.

Heffernan, William D., Mary Hendrickson, and Robert Gronski. 1999. *Consolidation in the Food and Agriculture System*. Report to the National Farmers Union. Columbia, MO: Department of Sociology, University of Missouri.

Heffernan, William D., and Mary K. Hendrickson. 2002. "Multi-National Concentrated Food Processing and Marketing Systems and the Farm Crisis." Presented at Annual Meeting of the American Association for the Advancement of Science Symposium: Science and Sustainability, Boston, MA, February.

Heimes, Rita S. 2010. "Post-Sale Restrictions on Patented Seeds: Which Law Governs." *Wake Forest Intellectual Property Law Journal* 10 (2): 98–152.

Heinemann, Jack A., Melanie Massaro, Dorien S. Coray, Sarah Zanon Agapito-Tenfen, and Jiajun Dale Wen. 2014. "Sustainability and Innovation in Staple Crop Production in the US Midwest." *International Journal of Agricultural Sustainability* 12 (1): 71–88.

Heller, Matthew. 2014. "7-Eleven Takes Big Gulp Out Of American Dream." *MintPress News*, July 22. http://bit.ly/1rKwtre.

Helliker, Kevin. 2002. "In Natural Foods, a Big Name's No Big Help." *Wall Street Journal*, June 7, sec. B1.

Hendrickson, Mary K., and William D. Heffernan. 2002. "Opening Spaces through Relocalization: Locating Potential Resistance in the Weaknesses of the Global Food System." *Sociologia Ruralis* 42 (4): 347–69.

Hendrickson, Mary, and William Heffernan. 2003. "Lessons for Public Breeding from Structural Changes in the Agricultural Marketplace." In *Summit on Seeds and Breeds for 21st Century Agriculture*, edited by Michael Sligh and Laura Lauffer, 11–22. Pittsboro, NC: Rural Advancement Foundation International-USA.

Hendrickson, Mary, William D. Heffernan, Philip H. Howard, and Judith B. Heffernan. 2001. "Consolidation in Food Retailing and Dairy." *British Food Journal* 103 (10): 715–28.

Hendrickson, Mary, William Heffernan, David Lind, and Elizabeth Barham. 2008a. "Contractual Integration in Agriculture: Is There a Bright Side for Agriculture of the Middle?" In *Food and the Mid-Level Farm: Renewing an Agriculture of the Middle*, edited by Thomas A. Lyson, G. W. Stevenson, and Rick Welsh, 79–100. Cambridge, MA: MIT Press.

Hendrickson, Mary, John Wilkinson, William D. Heffernan, and Robert Gronski. 2008b. *The Global Food System and Nodes of Power*. Boston, MA: Oxfam America. http:// papers.ssrn.com/abstract=1337273.

Henriques, Diana B. 1993. "Evidence Mounts of Rigged Bidding in Milk Industry." *The New York Times*, May 23. http://nyti.ms/Zm7OQO.

Henry, O. 1904. *Cabbages and Kings*. New York: Doubleday, Page & Company.

Hertz, Thomas, and Steven Zahniser. 2013. "Immigration and the Rural Workforce." *USDA Economic Research Service*. http://www.ers.usda.gov/topics/in-the-news/immigra-tion-and-the-rural-workforce.aspx.

Hiland Dairy, Inc. v Kroger Company. 1968. US Court of Appeals 8th Circuit.

Hilbeck, Angelika, Tamara Lebrecht, Raphaela Vogel, Jack A. Heinemann, and Rosa Binimelis. 2013. "Farmer's Choice of Seeds in Four EU Countries Under Different Levels of GM Crop Adoption." *Environmental Sciences Europe* 25 (1): 1–13.

Hildebrandt, Stephanie. 2012. "Wholesaler of the Year: Silver Eagle Distributors." *Beverage Industry*, September. http://digital.bnpmedia.com/display_article.php?id=1162037.

Hindy, Steve. 2014. "Free Craft Beer!" *The New York Times*, March 29. http://nyti. ms/1ktotXh.

Hindy, Steve, and Tom Potter. 2011. *Beer School: Bottling Success at the Brooklyn Brewery*. Hoboken, NJ: John Wiley & Sons.

Hoffman, Beth. 2013. "U.S. Hispanics Buy More Groceries At Convenience Stores Than Non-Hispanics, Study Finds." *Forbes*, July 29. http://onforb.es/XlqONS.

Holmes, Thomas J. 2011. "The Diffusion of Wal-Mart and Economies of Density." *Econometrica* 79 (1): 253–302.

Holt, Douglas B. 2003. "Interview with Steven Jackson." *Advertising & Society Review* 4 (2). doi:10.1353/asr.2003.0010, http://muse.jhu.edu/journals/advertising_and_soci-ety_review/summary/v004/4.2holt.html.

Hornick, Mike. 2013. "Ready Pac Dominates Single-Serve." *The Packer*, November 4. http://bit.ly/1qVBTj6.

Horovitz, Bruce. 2013. "7-Eleven Wants to Be Your Healthy-Snack Store." *USA Today*, September 12. http://usat.ly/1q5L5xt.

Horovitz, Bruce. 2014. "Fast-Food Worker Strike about to Go Global." *USA Today*, May 7. http://usat.ly/1D9fzsQ.

Horstmeier, Greg. 1996. "Strategic Bedfellows." *Farm Journal*, October, 18–19.

Howard, Phil, Terra Bogart, Alix Grabowski, Rebecca Mino, Nick Molen, and Steve Schultze. 2012. *Concentration in the U.S. Wine Industry*. East Lansing, MI, USA: Michigan State University. https://www.msu.edu/~howardp/wine.html.

Howard, Philip H. 2009a. "Visualizing Food System Concentration and Consolidation." *Southern Rural Sociology* 24 (2): 87–110.

Howard, Philip H. 2009b. "Consolidation in the North American Organic Food Processing Sector, 1997 to 2007." *International Journal of Sociology of Agriculture and Food* 16 (1): 13–30.

Howard, Philip H. 2009c. "Visualizing Consolidation in the Global Seed Industry." *Sustainability* 1 (4): 1266–87.

Howard, Philip H. 2011. "Increasing Community Participation with Self-Organizing Meeting Processes." *Journal of Rural Social Sciences* 27 (2): 118–36.

Howard, Philip H. 2014a. "Transnational Corporations." In *Achieving Sustainability: Visions, Principles and Practices*, edited by Debra Rowe, 737–42. Detroit, MI: Macmillan.

Howard, Philip H. 2014b. "Too Big to Ale? Globalization and Consolidation in the Beer Industry." In *The Geography of Beer: Regions, Environment, and Society*, edited by Mark W. Patterson and Nancy Hoast Pullen, 155–65. New York: Springer.

Howard, Philip H., and Patricia Allen. 2008. "Consumer Willingness to Pay for Domestic 'fair Trade': Evidence from the United States." *Renewable Agriculture and Food Systems* 23 (3): 235–42.

Howard, Philip H., and Patricia Allen. 2010. "Beyond Organic and Fair Trade? An Analysis of Ecolabel Preferences in the United States." *Rural Sociology* 75 (2): 244–69.

Howard, Philip H., Margaret Fitzpatrick, and Brian Fulfrost. 2011. "Proximity of Food Retailers to Schools and Rates of Overweight Ninth Grade Students: An Ecological Study in California." *BMC Public Health* 11: 68.

Howard, Philip H., and Daniel Jaffee. 2013. "Tensions Between Firm Size and Sustainability Goals: Fair Trade Coffee in the United States." *Sustainability* 5 (1): 72–89.

Hu, Janny. 2013. "Trader Joe's Frozen Oatmeal Wins Test." *San Francisco Chronicle*, January 4. http://bit.ly/1tTWCUh.

Hubbard, Kristina. 2009. *Out of Hand: Farmers Face the Consequences of a Consolidated Seed Industry*. Washington, DC: National Family Farm Coalition.

Huber, Bridget. 2014. "How to Take down a Terrible-Smelling Hog Farm." *Mother Jones*, June. http://bit.ly/1gqBa7T.

Huffstutter, P. J., and Lisa Baertlein. 2013. "Behind China's U.S. Pork Deal, Fears over Feed Additives." *Reuters*, May 30. http://reut.rs/1qVC1PG.

IBISWorld. 2013. *Fast Food Restaurants in the US*. Santa Monica, CA: IBISWorld.

Ingram, Paul, Lori Qingyuan Yue, and Hayagreeva Rao. 2010. "Trouble in Store: Probes, Protests, and Store Openings by Wal-Mart, 1998–2007." *American Journal of Sociology* 116 (1): 53–92.

Interbrand. 2013. "Best Global Brands 2013: Coca-Cola." http://www.interbrand.com/en/best-global-brands/2013/Coca-Cola.

Isakson, S. Ryan. 2014. "Food and Finance: The Financial Transformation of Agro-Food Supply Chains." *Journal of Peasant Studies* 41 (5): 749–775.

Isidore, Chris, and Jeanne Sahadi. 2014. "Burger King Buying Tim Hortons." *CNN Money*, August 27. http://cnnmon.ie/1viPHYY.

Izzo, Dina. 2007. "ALBA Organics." http://www.albafarmers.org/ALBA%20Organics.htm.

Jackson-Smith, Douglas B., and Frederick H. Buttel. 1998. "Explaining the Uneven Penetration of Industrialization in the U.S. Dairy Sector." *International Journal of Sociology of Agriculture and Food* 7: 113–50.

Jaffe, JoAnn, and Michael Gertler. 2006. "Victual Vicissitudes: Consumer Deskilling and the (Gendered) Transformation of Food Systems." *Agriculture and Human Values* 23 (2): 143–62.

Jaffee, Daniel. 2007. *Brewing Justice: Fair Trade Coffee, Sustainability, and Survival*. Berkeley, CA: University of California Press.

Jaffee, Daniel, and Philip H. Howard. 2010. "Corporate Cooptation of Organic and Fair Trade Standards." *Agriculture and Human Values* 27 (4): 387–99.

James, Harvey S., Jr., Mary Hendrickson, and Philip H. Howard. 2013. "Networks, Power and Dependency in the Agrifood Industry." In *The Ethics and Economics of Agrifood Competition*, edited by Harvey S. James Jr., 99–126. New York: Springer.

Jayaraman, Saru. 2014. *Shelved: How Wages and Working Conditions for California's Food Retail Workers Have Declined as the Industry Has Thrived*. Berkeley, CA: Food Labor Research Center, University of California.

Jernigan, David H. 2009. "The Global Alcohol Industry: An Overview." *Addiction* 104 (s1): 6–12.

Jia, Panle. 2008. "What Happens When Wal-Mart Comes to Town: An Empirical Analysis of the Discount Retailing Industry." *Econometrica* 76 (6): 1263–1316.

Johnson, Nancy L., and Vernon W. Ruttan. 1994. "Why Are Farms So Small?" *World Development* 22 (5): 691–706.

Johnson, Nathanael. 2006. "Swine of the Times." *Harper's Magazine*, May, 47–56.

Johnston, Josée. 2008. "The Citizen-Consumer Hybrid: Ideological Tensions and the Case of Whole Foods Market." *Theory and Society* 37 (3): 229–70.

Johnston, Josée, and Kate Cairns. 2013. "Searching for the Alternative, Caring, Reflexive Consumer." *International Journal of Sociology of Agriculture and Food* 20 (3): 403–8.

Jowit, Juliette. 2009. "Howard-Yana Shapiro: Man from Mars Trying to Save the Planet." *The Guardian*, May 7. http://bit.ly/1y4wYmk.

Kaiman, Jonathan. 2013. "China's Fast-Food Pioneer Struggles to Keep Customers Saying 'YUM!'" *The Guardian*, January 4. http://bit.ly/YK7YRW.

Kaplow, Louis. 1985. "Extension of Monopoly Power Through Leverage." *Columbia Law Review* 85 (3): 515–56.

Karp, Robert, and Thea Maria Carlson. 2014. "Deep Organic: An Intro to Biodynamics." *Dark Rye*, April. http://www.darkrye.com/content/deep-organic-intro-biodynamics.

Kastler, Guy. 2005. "Seed Laws in Europe: Locking Farmers Out." *Seedling*, July. http://www.grain.org/article/entries/541-seed-laws-in-europe-locking-farmers-out.

Kaufman, Phil. 2002. "Food Retailing." In *The U.S. Food Marketing System, 2002: Competition, Coordination, and Technological Innovations into the 21st Century*, edited by Steve W. Martinez, 21–33. Washington, DC: USDA Economic Research Service.

Kautsky, Karl. 1988. *The Agrarian Question*. Vol. 1–2. Winchester, MA: Zwan Publications.

Kelleher, Kevin. 2004. "66,207,896 Bottles of Beer on the Wall." *CNN.com*, February 25. http://cnn.it/1uD31oD.

Kelley, Christopher R. 2003. *An Overview of the Packers and Stockyards Act*. Fayetteville, AR: The National Agricultural Law Center, University of Arkansas. http://new.national-aglawcenter.org/wp-content/uploads/assets/articles/kelley_packers.pdf.

Kendall, Brent, and Valerie Bauerlein. 2013. "U.S. Sues to Block Big Beer Merger." *Wall Street Journal*, January 31. http://on.wsj.com/1lNmHPB.

Kennedy, Robert, Jr. 2003. "Smithfield Foods: The Truth behind Its Pigs and Factories." *The Ecologist*, December 1. http://bit.ly/1q5LzDY.

Keupper, George. 2010. *A Brief Overview of the History and Philosophy of Organic Agriculture*. Poteau, OK: Kerr Center for Sustainable Agriculture.

Key, Nigel D., and Michael J. Roberts. 2007. "Do Government Payments Influence Farm Size and Survival?" *Journal of Agricultural and Resource Economics* 32 (2): 330–48.

Khan, Lina. 2013a. "How Monsanto Outfoxed the Obama Administration." *Salon*, March 15. http://bit.ly/1BEYwx6.

Khan, Lina. 2013b. "The Folks Who Sell Your Corn Flakes Are Acting Like Goldman Sachs—and That Should Worry You." *The New Republic*, September 11. http://bit.ly/1qCt6Vw.

Khan, Lina. 2013c. "Monsanto's Scary New Scheme: Why Does It Really Want All This Data?" *Salon*, December 29. http://bit.ly/1cAZEqs.

Khoury, Colin K., Anne D. Bjorkman, Hannes Dempewolf, Julian Ramirez-Villegas, Luigi Guarino, Andy Jarvis, Loren H. Rieseberg, and Paul C. Struik. 2014. "Increasing Homogeneity in Global Food Supplies and the Implications for Food Security." *Proceedings of the National Academy of Sciences* 111 (11): 4001–4006.

Kleiner, Anna M., and John J. Green. 2009. "The Contributions of Dr. William Heffernan and the Missouri School of Agrifood Studies." *Southern Rural Sociology* 24 (2): 14–28.

Kloppenburg, Jack. 2004. *First the Seed: The Political Economy of Plant Biotechnology*. Madison, WI: University of Wisconsin Press.

Kloppenburg, Jack. 2010. "Impeding Dispossession, Enabling Repossession: Biological Open Source and the Recovery of Seed Sovereignty." *Journal of Agrarian Change* 10 (3): 367–88.

Kloppenburg, Jack. 2014. "Re-Purposing the Master's Tools: The Open Source Seed Initiative and the Struggle for Seed Sovereignty." *Journal of Peasant Studies* 41 (6): 1225–1246.

Kneen, Brewster. 1993. *From Land to Mouth: Understanding the Food System*. Toronto, ON, Canada: NC Press.

Kneen, Brewster. 2013. "Disconnecting the Dots: Boundaries and Rights." Presented at the Yale Agrarian Studies Colloquium, New Haven, CT, September 13.

Knoedelseder, William. 2012. *Bitter Brew: The Rise and Fall of Anheuser-Busch and America's Kings of Beer*. New York: HarperCollins.

Kolko, Gabriel. 1963. *The Triumph of Conservatism*. New York: The Free Press.

Kowitt, Beth. 2014. "Whole Foods Takes Over America." *Fortune Magazine*. http://fortune.com/2014/04/10/whole-foods-takes-over-america/.

Kozloff, Nikolas. 2009. "From Arbenz to Zelaya: Chiquita in Latin America." *Counterpunch*, July 17. http://www.counterpunch.org/2009/07/17/chiquita-in-latin-america/.

Krebs, Valdis, and June Holley. 2006. *Building Smart Communities Through Network Weaving*. Athens, OH: The Appalachian Center for Economic Networks, Inc. (ACEnet).

Kroger, Chris. 2011. "This Generation Mixes It Up a Little More with Packaged Salads." *The Packer*, October 27. http://bit.ly/1qXnnsP.

Krulwich, Robert. 1996. "So You Want to Buy a President?" *Frontline*. PBS. http://www.pbs.org/wgbh/pages/frontline/president/players/andreas.html.

Kurlansky, Mark. 2014. "Inside the Milk Machine: How Modern Dairy Works." *Modern Farmer*, March 17. http://modernfarmer.com/2014/03/real-talk-milk/.

Kurtuluş, Mümin, Sezer Ülkü, Jeffrey P. Dotson, and Alper Nakkas. 2014. "The Impact of Category Captainship on the Breadth and Appeal of a Retailer's Assortment." *Journal of Retailing* 90 (3): 379–392.

Kutka, Frank. 2011. "Open-Pollinated vs. Hybrid Maize Cultivars." *Sustainability* 3 (9): 1531–54.

Lardner, George, Jr. 1974. "Anatomy of a Scandal: The Milk Fund." *Reader's Digest*, November.

Larkin, Judy. 2003. *Strategic Reputation Risk Management*. New York: Palgrave Macmillan.

Larson, Nicole I., Mary T. Story, and Melissa C. Nelson. 2009. "Neighborhood Environments: Disparities in Access to Healthy Foods in the U.S." *American Journal of Preventive Medicine* 36 (1): 74–81.

Lawrence, John D. 2010. "Hog Marketing Practices and Competition Questions." *Choices* 25 (2). http://www.choicesmagazine.org/magazine/article.php?article=122.

Lawrence, Robyn Griggs. 2011. "Really Big and Off the Grid." *Mother Earth News*, May 6. http://www.motherearthnews.com/green-homes/really-big-off-the-grid.aspx.

Lazich, Robert S., and Virgil L. Burton. 2014. *Top Refrigerated Kefir/Milk Substitute/Soy Milk Brands, 2012*. Market Share Reporter. Detroit, MI: Gale Group.

Leonard, Christopher. 2014. *The Meat Racket: The Secret Takeover of America's Food Business*. New York: Simon and Schuster.

Leonard, Devin. 2014. "Burger King is Run by Children." *BusinessWeek*, July 24. http://buswk.co/1rOm4ef.

Leschin-Hoar, Clare. 2011. "Feds Help GMO Salmon Swim Upstream." *Grist*. http://grist.org/food/2011-09-29-feds-help-gmo-salmon-swim-upstream/.

Levenstein, Harvey A. 1988. *Revolution at the Table: The Transformation of the American Diet*. Berkeley, CA: University of California Press.

Levenstein, Harvey A. 2003. *Paradox of Plenty: A Social History of Eating in Modern America*. Berkeley, CA: University of California Press.

Levidow, Les, and Susan Carr. 2007. "GM Crops on Trial: Technological Development as a Real-World Experiment." *Futures* 39 (4): 408–31.

Lewis, Truman. 2008. "Feds Probe Food Prices." *ConsumerAffairs*. http://www.consumer-affairs.com/news04/2008/09/feds_food.html.

Lewontin, Richard C., and Jean-Pierre Berlan. 1986. "Technology, Research, and the Penetration of Capital: The Case of U.S. Agriculture." *Monthly Review* 38(3): 21–34.

Lewontin, Richard C. 2000. "The Maturing of Capitalist Agriculture: Farmer as Proletarian." In *Hungry for Profit: The Agribusiness Threat to Farmers, Food, and the Environment*, edited by Fred Magdoff, John Bellamy Foster, and Frederick H. Buttel, 93–106. New York: Monthly Review Press.

Li, Jingsong, Edith T. Janice Jiggins, Lammerts Van Bueren, and Cees Leeuwis. 2013. "Towards a Regime Change in the Organization of the Seed Supply System in China." *Experimental Agriculture* 49 (1): 114–33.

Lieber, James B. 2000. *Rats in the Grain: The Dirty Tricks and Trials of Archer Daniels Midland, the Supermarket to the World*. New York: Four Walls Eight Windows.

Lim, Daryl. 2013. "Self-Replicating Technologies and the Challenge for the Patent and Antitrust Laws." *Cardozo Arts & Entertainment Law Journal* 32 (1): 131–223.

Lin, Fang. 2013. "Why Glyphosate Prices Are Rising." *Farm Chemicals International*, September 5. http://bit.ly/Zm8eGQ.

Linden, Tim. 2013. "FMSA Regulations: Produce Rule and Preventive Control Rule Receive Scrutiny." *Western Growers Magazine*, November 1. http://bit.ly/1tV5bSB.

Lipschutz, Ronnie D. 2010. *Political Economy, Capitalism, and Popular Culture*. Lanham, MD: Rowman & Littlefield.

Liston, Barbara. 2014. "Craft Beer Distribution Battle Brews in Florida Legislature." *Reuters*, April 4. http://reut.rs/Zm8fdO.

Lobao, Linda, and Curtis W. Stofferahn. 2008. "The Community Effects of Industrialized Farming: Social Science Research and Challenges to Corporate Farming Laws." *Agriculture and Human Values* 25 (2): 219–40.

Lobb, Richard. 2013. "Plant Biotechnology Companies Begin New Conversation About GMOs and How Our Food Is Grown." http://bit.ly/1BEYXaJ.

Lockie, Stewart, and Darren Halpin. 2005. "The 'Conventionalisation' Thesis Reconsidered: Structural and Ideological Transformation of Australian Organic Agriculture." *Sociologia Ruralis* 45 (4): 284–307.

Lockie, Stewart, and Simon Kitto. 2000. "Beyond the Farm Gate: Production-Consumption Networks and Agri-Food Research." *Sociologia Ruralis* 40 (1): 3–19.

Logan, Tim. 2012. "Beer Battle Between Wholesalers, Brewers." *St. Louis Post-Dispatch*, May 6. http://bit.ly/1q5MhB8.

Lynn, Barry C. 2006. "Breaking the Chain." *Harper's Magazine*, July. http://harpers.org/archive/2006/07/breaking-the-chain/.

Lyson, Thomas A., and Annalisa Lewis Raymer. 2000. "Stalking the Wily Multinational: Power and Control in the US Food System." *Agriculture and Human Values* 17 (2): 199–208.

MacDonald, James M., Penni Korb, and Robert A. Hoppe. 2013. *Farm Size and the Organization of U.S. Crop Farming*. Economic Research Report 152. Washington, DC: USDA Economic Research Service.

MacIntosh, Julie. 2011. *Dethroning the King: The Hostile Takeover of Anheuser-Busch, an American Icon*. Hoboken, NJ: John Wiley & Sons.

MacPherson, Allie, Kory Mello, and Amber Rinehard. 2009. *Whole Foods/Wild Oats Merger: Sowing the Seeds for Market Growth*. New York: Arthur W. Page Society.

Magdoff, Fred, and John Bellamy Foster. 2011. *What Every Environmentalist Needs to Know About Capitalism: A Citizen's Guide to Capitalism and the Environment*. New York: Monthly Review Press.

Magdoff, Fred, and Brian Tokar. 2010. "Agriculture and Food in Crisis: An Overview." In *Agriculture and Food in Crisis: Conflict, Resistance, and Renewal*, edited by Fred Magdof and Brian Tokar, 9–30. New York: NYU Press.

Magdoff, Harry, and Paul M. Sweezy. 1987. *Stagnation and the Financial Explosion*. New York: Monthly Review Press.

Makki, Shiva S., and Charles Plummer. 2005. "Globalization of the Frozen Potato Industry." *Journal of Agribusiness* 23 (2): 133–46.

Maltby, Ed. 2012. "Organic Dairy Pay Price Update." http://www.nodpa.com/payprice_update_01192012.shtml.

Mann, Susan A., and James M. Dickinson. 1978. "Obstacles to the Development of a Capitalist Agriculture." *Journal of Peasant Studies* 5 (4): 466–81.

Marion, Bruce W., and James M. MacDonald. 2013. "The Agriculture Industry." In *The Structure of American Industry*, edited by James Brock, Twelfth Edition, 1–29. Long Grove, IL: Waveland Press.

Martin, Andrew. 2012. "Awash in Milk and Money." *The New York Times*, October 28, BU1.

Martinez, Steve W. 2002. "Food Wholesaling." In *The U.S. Food Marketing System, 2002: Competition, Coordination, and Technological Innovations into the 21st Century*, edited by Steve W. Martinez, 12–20. Washington, DC: USDA Economic Research Service.

Martinez, Steve W. 2007. *The U.S. Food Marketing System: Recent Developments, 1997–2006*. Washington, DC: USDA Economic Research Service.

Martrenchar, A. 1999. "Animal Welfare and Intensive Production of Turkey Broilers." *World's Poultry Science Journal* 55 (2): 143–52.

Mascarenhas, Michael, and Lawrence Busch. 2006. "Seeds of Change: Intellectual Property Rights, Genetically Modified Soybeans and Seed Saving in the United States." *Sociologia Ruralis* 46 (2): 122–38.

Mattera, Philip. 2004. *USDA Inc.: How Agribusiness Has Hijacked Regulatory Policy at the U.S. Department of Agriculture*. Washington, DC: Corporate Research Project of Good Jobs First.

Maurer, Virginia G. 2009. "Corporate Social Responsibility and the 'Divided Corporate Self': The Case of Chiquita in Colombia." *Journal of Business Ethics* 88: 595–603.

McBride, William D., and Nigel Key. 2013. *U.S. Hog Production from 1992 to 2009: Commodity Payments, Farm Business Survival, and Farm Size Growth*. Economic Research Report 158. Washington, DC: USDA Economic Research Service.

McConnell, William. 2014a. "Antitrust Watchdog Wary of Sysco-US Foods Deal." *The Deal Pipeline*, February 25. http://bit.ly/YK8XBr.

McConnell, William. 2014b. "Some Independents Eager to Take Bite Out of Sysco." *The Deal Pipeline*, April 3. http://bit.ly/1qCuDLl.

McEvoy, Miles. 2014. "The Growth of an Industry – 18,000+ Certified Organic Operations in the U.S. Alone." *USDA Blog*. http://1.usa.gov/1eWPO0d.

McGrath, Maria. 2004. "That's Capitalism, Not a Co-Op: Countercultural Idealism and Business Realism in 1970s U.S. Food Co-Ops." *Business and Economic History On-Line* 2: 1–14. http://www.thebhc.org/publications/BEHonline/2004/McGrath.pdf.

McIntyre, Robert S., Matthew Gardner, Rebecca J. Wilkins, and Richard Phillips. 2011. *Corporate Tax Payers & Corporate Tax Dodgers*. Washington, DC: Citizens for Tax Justice & The Institute on Taxation and Economic Policy.

McNeil, Donald G., Jr. 2015. "A Flu Epidemic That Threatens Birds, Not Humans." *The New York Times*, May 4. http://nyti.ms/1JOjOPc.

Meek, Andy. 2006. "Down and Out in Covington." *Memphis Daily News*, June 22. http://www.memphisdailynews.com/editorial/Article.aspx?id=30496.

Melcarek, Hilary, and Patricia Allen. 2013. *The Human Face of Sustainable Food Systems: Adding People to the Environmental Agenda*. Center for Agroecology and Sustainable Food Systems, University of California Santa Cruz.

Mertes, Micah. 2013. "Happy 20th Birthday Demolition Man!" *Omaha World-Herald*, October 8. http://bit.ly/1s2Klvb.

Mintel. 2013a. *Dairy and Non-Dairy Milk, US*. London, UK: Mintel.

Mintel. 2013b. *Fruit and Vegetables, US*. London, UK: Mintel.

Mintel. 2013c. *Beer, US*. London, UK: Mintel.

Mitchell, Stacy. 2012. *Walmart's Greenwash*. Minneapolis, MN: Institute for Local Self-Reliance.

Moeller, David R. 2003. "The Problem of Agricultural Concentration: The Case of the Tyson-IBP Merger." *Drake Journal of Agricultural Law* 8: 1–24.

Monsanto. 2014. "Monsanto's Seeds and Traits Performance Drives Strong Second Quarter Growth, Reinforces Confidence in Full-Year Outlook." http://bit.ly/1khmBVN.

Moodie, Rob, David Stuckler, Carlos Monteiro, Nick Sheron, Bruce Neal, Thaksaphon Thamarangsi, Paul Lincoln, and Sally Casswell. 2013. "Profits and Pandemics: Prevention of Harmful Effects of Tobacco, Alcohol, and Ultra-Processed Food and Drink Industries." *The Lancet* 381 (9867): 670–79.

Mooney, Patrick H. 2004. "Democratizing Rural Economy: Institutional Friction, Sustainable Struggle and the Cooperative Movement." *Rural Sociology* 69 (1): 76–98.

Moretti, Irene Musselli. 2006. *Tracking the Trend Towards Market Concentration: The Case of the Agricultural Input Industry*. Geneva, Switzerland: United Nations Conference on Trade and Development.

Morgan, Dan. 1979. *Merchants of Grain*. New York: Viking Press.

Moss, Michael. 2013. *Salt Sugar Fat: How the Food Giants Hooked Us*. New York: Random House.

Mount, Jeffrey, Emma Freeman, and Jay Lund. 2014. *Water Use in California*. San Francisco, CA: Public Policy Institute of California.

MSI. 2007. "Frequently Asked Questions." http://www.msilimited.com/downloads/faq.pdf.

Munck, Ronaldo. 2006. "Globalization and Contestation: A Polanyian Problematic." *Globalizations* 3 (2): 175–86.

Murphy, Sophia. 2006. *Concentrated Market Power and Agricultural Trade*. Berlin: Heinrich Böll Foundation.

Murphy, Sophia. 2009. "Free Trade in Agriculture: A Bad Idea Whose Time Is Done." *Monthly Review* 61 (3). http://bit.ly/1D9hptG.

Murphy, Sophia, David Burch, and Jennifer Clapp. 2012. *Cereal Secrets: The World's Largest Grain Traders and Global Agriculture*. Oxfam Research Reports. Oxford, UK: Oxfam.

Murray, Douglas W., and Martin A. O'Neill. 2012. "Craft Beer: Penetrating a Niche Market." *British Food Journal* 114 (7): 899–909.

National Corn Growers Association. 2014. *World of Corn 2014*. Chesterfield, MO: National Corn Growers Association. http://www.ncga.com/upload/files/documents/pdf/woc-2014.pdf.

National Farmers Union. 2014a. *The Price of Patented Seed, The Value of Farm Saved Seed*. Saskatoon, SK, Canada.

National Farmers Union. 2014b. *Farmer's Share of Retail Food Dollar*. Washington, DC: National Farmers Union. http://nfu.org/images/082814_FarmersShare.pdf.

National Pork Board. 2014. "Pork Checkoff: Research." http://www.pork.org/Research/Default.aspx.

National Sustainable Agriculture Coalition. 2014. "What Is the GIPSA Rider and Why Is the House Once Again Attacking Farmers' Rights?" http://sustainableagriculture.net/blog/what-is-the-gipsa-rider/.

Natzke, Dave. 2013. "DFA Settles Cheese Trading Lawsuit." *Dairy Business*, March 22. http://bit.ly/1uD4pYl.

NBWA. 2014. *Benefits of Beer Franchise Laws*. Alexandria, VA: National Beer Wholesalers Association. http://nbwa.org/sites/default/files/Benefits_of_Beer_Franchise_Laws_Brochure.pdf.

Neff, Roni A., Anne M. Palmer, Shawn E. Mckenzie, and Robert S. Lawrence. 2009. "Food Systems and Public Health Disparities." *Journal of Hunger & Environmental Nutrition* 4 (3–4): 282–314.

Nestle, Marion. 2014. "The Farm Bill Promotes Fruits and Vegetables? Really?" *Food Politics*. http://www.foodpolitics.com/2014/03/the-farm-bill-promotes-fruits-and-vegetables-really/.

Neubauer, Chuck. 2009. "Monsanto Chief Accuses Rival DuPont of Deceit." *The Washington Times*, August 18. http://bit.ly/1uyY8xN.

Neuburger, Bruce. 2013. *Lettuce Wars*. New York: NYU Press.

Neuman, W. Lawrence. 1998. "Negotiated Meanings and State Transformation: The Trust Issue in the Progressive Era." *Social Problems* 45 (3): 315–35.

New America Foundation. 2012. *A King of Beers?* Washington, DC: Markets, Enterprise and Resiliency Initiative.

Nguyen, Vicky, Kevin Nious, Felipe Escamilla, David Paredes, Mark Villarreal, and Jeremy Carroll. 2013. "Food in Dirty Sheds Served to Bay Area Restaurants." *NBC Bay Area*, October 14. http://bit.ly/1qCvIlo.

Nicklaus, David. 2015. "Syngenta Deal Would Bulk Up Monsanto in Crop Chemicals." *St. Louis Post-Dispatch*, May 5. http://bit.ly/1GMvrAb.

Nicks, Denver. 2013. "Apparently McDonald's Thinks Its Employees Have Swimming Pools and Nannies." *Time*, December 6. http://ti.me/II8sPo.

Nitzan, Jonathan. 1998. "Differential Accumulation: Towards a New Political Economy of Capital." *Review of International Political Economy* 5 (2): 169–216.

Nitzan, Jonathan, and Shimshon Bichler. 2009. *Capital as Power: A Study of Order and Creorder*. New York: Routledge.

Nitzan, Jonathan, and Shimshon Bichler. 2014. "Profit from Crisis." *Frontline*, May 2. http://www.frontline.in/world-affairs/profit-from-crisis/article5915462.ece.

O'Brien, Doug. 2004. "Policy Approaches to Address Problems Associated with Consolidation and Vertical Integration in Agriculture." *Drake Journal of Agricultural Law* 9: 33–52.

O'Connor, James. 1973. *Fiscal Crisis of the State*. New York: St. Martin's Press.

O'Halloran, Shane. 2014. "White Wave Plans Earthbound Farm Expansion." *Food Engineering*, March 12. http://bit.ly/1uObapm.

Oates, Bryce. 2000. "Taking on Corporate Pork." *Multinational Monitor*. http://www.multinationalmonitor.org/mm2000/072000/interview-oates.html.

Obach, Brian K. 2015. *Organic Struggle: The Movement for Sustainable Agriculture in the United States*. Cambridge, MA: MIT Press.

Ogburn, Stephanie Paige. 2011. "Cattlemen Struggle Against Giant Meatpackers and Economic Squeezes." *High Country News*, March 21. http://bit.ly/1sUvG90.

Olympia Food Co-op. 2013. "Co-Op Resumes Ordering from UNFI - Teamsters Strike Continues." http://bit.ly/YK9ByV.

Olender, Katie. 2007. "Delivering on a Quality Promise." *New Farm*, December 7. http://newfarm.rodaleinstitute.org/features/2007/1207/fooddesert/olender.shtml.

Ollinger, Michael, Sang V. Nguyen, Donald P. Blayney, William Chambers, and Kenneth B. Nelson. 2005. *Structural Change in the Meat, Poultry, Dairy and Grain Processing Industries*. Economic Research Report 7217. Washington, DC: United States Department of Agriculture, Economic Research Service.

Olson, R. Dennis. 2014. "Lessons from the Food System: Borkian Paradoxes, Plutocracy, and the Rise of Walmart's Buyer Power." In *The Global Food System:*

Issues and Solutions, edited by William D. Schanbacher, 83–114. Santa Barbara, CA: ABC-CLIO.

Ordman, Marty. 2012. *Dole Fresh Vegetables Announces Precautionary Recall of Limited Number of Salads*. Washington, DC: U.S. Food and Drug Administration. http://www.fda.gov/Safety/Recalls/ucm309601.htm.

Orgel, David, David Merrefield, and David Ghitelman. 1999. "The Deal's Off: Ahold Abandons Pathmark Buyout." *Supermarket News*, December 20. http://supermarket-news.com/archive/deals.

Otero, Gerardo, Gabriela Pechlaner, and Efe Can Gürcan. 2013. "The Political Economy of Food Security and Trade: Uneven and Combined Dependency." *Rural Sociology* 78 (3): 263–89.

Otten, Jennifer J., Brian E. Saelens, Kristopher I. Kapphahn, Eric B. Hekler, Matthew P. Buman, Benjamin A. Goldstein, Rebecca A. Krukowski, Laura S. O'Donohue, Christopher D. Gardner, and Abby C. King. 2014. "Impact of San Francisco's Toy Ordinance on Restaurants and Children's Food Purchases, 2011–2012." *Preventing Chronic Disease* 11: 14026.

Pachirat, Timothy. 2011. *Every Twelve Seconds*. New Haven, CT: Yale University Press.

Page, Greg. 2008. "Trusting Photosynthesis." Presented at the Chautauqua Institution, Chautauqua, NY, August. http://www.cargill.com/news/speeches-presentations/trusting-photosynthesis/index.jsp.

Palmer, Randall, and Allison Martell. 2014. "Canada Court Rules Against Wal-Mart over Quebec Store Closure." *Reuters*, June 27. http://reut.rs/1qHBlAg.

Parker, Russell C. 1976. *The Status of Competition in the Food Manufacturing and Food Retailing Industries*. Working Paper 6. Madison, WI: University of Wisconsin.

Paul, Katherine. 2013. "Victory for Fair Trade and Food Workers' Rights." *Organic Consumers Association*. http://www.organicconsumers.org/articles/article_27025.cfm.

Pearce, Fred. 2002. "How the Other Half Lives." *New Scientist*, August 3.

Peekhaus, Wilhelm C. 2013. *Resistance Is Fertile: Canadian Struggles on the Biocommons*. Vancouver, BC: University of British Columbia Press.

Peña, Nickie. 2013. "Maui Brewing to Join Stone Brewing Co. in Distributing Craft Beer Brands in Maui." *Craft Beer*, February 13. http://bit.ly/1uyYnJl.

Perelman, Michael. 2000. *The Invention of Capitalism: Classical Political Economy and the Secret History of Primitive Accumulation*. Durham, NC: Duke University Press.

Performance Food Group. 2012. "Performance Food Group Announces Agreement to Acquire Fox River Foods, Inc." December 12. http://bit.ly/1m4Nw8F.

Peter, Laurence. 2009. "Germans Protest over Pig Patent." *BBC*, April 16. http://bbc.in/1FEhd8d.

Peterson, E. Wesley F. 2009. *A Billion Dollars a Day: The Economics and Politics of Agricultural Subsidies*. Malden, MA: Wiley-Blackwell.

Peterson, Hayley. 2013. "A British Supermarket Chain Is Installing Creepy Face-Scanning Cameras to Track Consumers." *Business Insider*, November 5. http://read.bi/1uObggJ.

Peterson, Kim. 2012. "Budweiser Sees Big Sales Drop." *MSNMoney*, November 1. http://on-msn.com/1oATwDl.

Philpott, Tom. 2012. "DOJ Mysteriously Quits Monsanto Antitrust Investigation." *Mother Jones*. http://bit.ly/1nWsBjx.

Piketty, Thomas. 2014. *Capital in the Twenty-First Century*. Cambridge, MA: Harvard University Press.

Pilling, Dafydd, and Barbara Rischkowsky, eds. 2007. *The State of the World's Animal Genetic Resources for Food and Agriculture*. Rome, Italy: United Nations Food and Agriculture Organization.

Pitt, David. 2013. "DuPont Wins Fight for South African Seed Company." *Associated Press*, July 31. http://yhoo.it/1wuEUf1.

Pitt, David. 2014. "US Farmers Plant Record Soybean Crop, Less Corn." *Associated Press*, June 30. http://yhoo.it/1AP149E.

Plain, Ron. 2003. "Prediction of Pork Prices and Production." Presented at the Banff Pork Seminar, Banff, Alberta, Canada.

Polanyi, Karl. 1944. *The Great Transformation: The Political and Economic Origins of Our Time*. Boston, MA: Beacon Press.

Pols, Mary. 2014. "Maine Farmer, Seed Curator Forms New Grass-Roots Group." *Portland Press Herald*, February 16. http://www.pressherald.com/news/The_Maine_man_who_saves_the_seeds_.html.

Postel, Sandra L. 2000. "Entering an Era of Water Scarcity: The Challenges Ahead." *Ecological Applications* 10 (4): 941–48.

Potts, Monica. 2011. "The Serfs of Arkansas." *The American Prospect*, March 5. http://prospect.org/article/serfs-arkansas-0.

Pous, Terri. 2012. "Pre-Peeled Bananas Incur the Wrath of Humanity." *Time*, September 26. http://ti.me/1bx1gjx.

Powell, Lisa M., M. Christopher Auld, Frank J. Chaloupka, Patrick M. O'Malley, and Lloyd D. Johnston. 2007. "Associations Between Access to Food Stores and Adolescent Body Mass Index." *American Journal of Preventive Medicine* 33 (4): S301–7.

Preville, Philip. 2014. "How Couche-Tard Conquered the World, Starting with a Single Dépanneur." *Canadian Business*, June 3. http://bit.ly/1uyYCE2.

Pritchard, Bill, and David Burch. 2003. *Agri-Food Globalization in Perspective: International Restructuring in the Processing Tomato Industry*. Farnham, UK: Ashgate Publishing.

Progressive Grocer. 2002. "PG Profile: Kroger." *Progressive Grocer*, November 14. http://www.progressivegrocer.com/node/65603.

Provost, Claire, and Felicity Lawrence. 2012. "US Food Aid Programme Criticised as 'Corporate Welfare' for Grain Giants." *The Guardian*, July 18. http://bit.ly/1wtX29Q.

Public Citizen. 2004. *Smithfield Foods: A Corporate Profile*. Washington, DC: Public Citizen. http://www.citizen.org/documents/Smithfield.pdf.

Quigley, Kelly. 2003. "Milk Price-Fixing Suit Dismissed." *Crain's Chicago Business*, February 25. http://bit.ly/1s2Lk40.

Quivira. 2014. "Biodynamic Farming." http://www.quivirawine.com/index.php?option= com_submenus&id=2.

Ratner, Jonathan. 2014. "Alimentation Couche-Tard May Look Outside U.S. or to Supermarkets for Next Deal." *National Post*, June 25. http://bit.ly/1rxQxvT.

Ray, Daryll E., and Harwood D. Schaffer. 2012. "Public Trust is a Terrible Thing to Waste." *Policy Pennings*. http://agpolicy.org/weekcol/619.html.

Ray, Daryll E., Daniel de la Torre Ugarte, and Kelly Tiller. 2003. *Rethinking US Agricultural Policy*. Knoxville, TN: Agriculture Policy Analysis Center, University of Tennessee. http:// agris.fao.org/agris-search/search.do?recordID=US201300091343.

Renard, Marie-Christine. 2003. "Fair Trade: Quality, Market and Conventions." *Journal of Rural Studies* 19 (1): 87–96.

Renaud, Erica N. C., Edith T. Lammerts van Bueren, and Janice Jiggins. 2014. "The Development and Implementation of Organic Seed Regulation in the USA." *Organic Agriculture* 4(1): 25–42.

Richards, Carol, Hilde Bjørkhaug, Geoffrey Lawrence, and Emmy Hickman. 2013. "Retailer-Driven Agricultural restructuring—Australia, the UK and Norway in Comparison." *Agriculture and Human Values* 30 (2): 235–45.

Riedl, Brian M. 2007. *How Farm Subsidies Harm Taxpayers, Consumers, and Farmers, Too*. Washington, DC: The Heritage Foundation.

Ritzer, George. 2013. *The McDonaldization of Society: 20th Anniversary Edition*. Thousand Oaks, CA: SAGE.

Robb, Walter, and A. C. Gallo. 2013. "GMO Labeling Coming to Whole Foods Market." *Whole Story*. http://bit.ly/1sUwZ84.

Robinson, Jo. 2013. *Eating on the Wild Side: The Missing Link to Optimum Health*. New York: Little, Brown & Co.

Rosegrant, Mark W., Claudia Ringler, and Tingju Zhu. 2009. "Water for Agriculture: Maintaining Food Security under Growing Scarcity." *Annual Review of Environment and Resources* 34 (1): 205–22.

Rosen, Steve. 2013. "Dairy Farmers to Get $158.6 Million in Price-Fixing Settlement." *Kansas City Star*, January 22. http://www.agweek.com/event/article/id/20421/.

Rosenbaum, Aliza, and Rob Cox. 2009. "Big Money: Is Big Beer Begging for an Antitrust Probe?" *The Washington Post*, September 6. http://wapo.st/1hjSwE7.

Rosenwald, Michael S. 2008. "A Stronger Brew?" *The Washington Post*, February 6. http:// wapo.st/1oh5Jjn.

Rozendaal, Esther, Matthew A. Lapierre, Eva A. van Reijmersdal, and Moniek Buijzen. 2011. "Reconsidering Advertising Literacy as a Defense Against Advertising Effects." *Media Psychology* 14 (4): 333–54.

Ruben, Dennis L. 2013. "C-Store Industry Year in Review." *CSP Daily News*, January 25. http://www.nrc.com/services/aboutus/news/articles/272.html.

Ruetschlin, Catherine. 2014. *Fast Food Failure: How CEO-to-Worker Pay Disparity Undermines the Industry and the Overall Economy*. New York: Demos.

Rumbo, Joseph D. 2002. "Consumer Resistance in a World of Advertising Clutter: The Case of Adbusters." *Psychology and Marketing* 19 (2): 127–48.

Russell, Michael. 2011. "Portland's Best New Food Carts." *The Oregonian*, June 24. http://bit.ly/1m4NVYP.

Salerno, Christopher. 2014. "Can Anything Stop WhiteWave Foods?" *Seeking Alpha*, May 22. http://seekingalpha.com/article/2232663-can-anything-stop-whitewave-foods.

SANA. 2014. *History of Soy Products*. Washington, DC: Soyfoods Association of North America. http://www.soyfoods.org/soy-products/history-of-soy-products.

Scanlan, Stephen J. 2013. "Feeding the Planet or Feeding Us a Line? Agribusiness, 'Grainwashing' and Hunger in the World Food System." *International Journal of Sociology of Agriculture and Food* 20 (3): 357–82.

Schafer, Sara. 2013. "Inside the Seed Industry." *Farm Journal*, July 25. http://www.agweb.com/article/inside_the_seed_industry/.

Scherer, Frederic M., and David Ross. 1990. *Industrial Market Structure and Economic Performance*. 3rd ed. Boston, MA: Houghton Mifflin Company.

Schimmelpfennig, David E., Carl E. Pray, and Margaret F. Brennan. 2004. "The Impact of Seed Industry Concentration on Innovation: A Study of US Biotech Market Leaders." *Agricultural Economics* 30 (2): 157–67.

Schlosser, Eric. 2001. *Fast Food Nation: The Dark Side of the All-American Meal*. New York: Houghton Mifflin Harcourt.

Schlozman, Kay Lehman, Sidney Verba, and Henry E. Brady. 2012. *The Unheavenly Chorus: Unequal Political Voice and the Broken Promise of American Democracy*. Princeton, NJ: Princeton University Press.

Schneiberg, Marc, Marissa King, and Thomas Smith. 2008. "Social Movements and Organizational Form: Cooperative Alternatives to Corporations in the American Insurance, Dairy, and Grain Industries." *American Sociological Review* 73 (4): 635–67.

Schnepf, Randy. 2014. *International Food Aid Programs: Background and Issues*. 7–5700. Washington, DC: Congressional Research Service.

Schneyer, Joshua. 2011. "Commodity Traders: The Trillion Dollar Club." *Reuters*, October 28. http://reut.rs/1qHtAZ6.

Schor, Juliet. 2004. *Born to Buy: The Commercialized Child and the New Consumer Culture*. New York: Scribner.

Schultz, Mark. 2014. "Counterpoint: If the Trans-Pacific Partnership Is So Swell, Why All the Secrecy?," May 22. http://www.startribune.com/opinion/commentaries/260338971.html.

Schumacher, Harry. 2014. "Top 30 US Beer Distributors," March 17. http://katzamericas.blogspot.com/2014/03/list-top-30-us-beer-distributors.html.

Schurman, Rachel, and William A. Munro. 2010. *Fighting for the Future of Food: Activists Versus Agribusiness in the Struggle Over Biotechnology*. Minneapolis, MN: University of Minnesota Press.

Scott, Mark. 2013. "Anheuser-Busch InBev Revises Twenty Point One Billion Takeover Plan." *New York Times*, February 14. http://nyti.ms/1tV7d5p/.

Sell, Susan K. 2003. *Private Power, Public Law: The Globalization of Intellectual Property Rights*. Boston, MA: Cambridge University Press.

Sen, Amartya. 1990. *Poverty and Famines: An Essay on Entitlement and Deprivation*. Oxford: Oxford University Press.

Seubsman, Sam-ang, Matthew Kelly, Pataraporn Yuthapornpinit, and Adrian Sleigh. 2009. "Cultural Resistance to Fast-Food Consumption? A Study of Youth in North Eastern Thailand." *International Journal of Consumer Studies* 33 (6): 669–75.

Sewell, Josh. 2014. *Shallow Loss Programs May Surpass Cost of Old Subsidy Schemes*. Washington, DC: Taxpayers for Common Sense.

Shackleton, Holly. 2014. "Booths Pay Highest Price for 'Fair Milk.'" *Speciality Food Magazine*, May 20. http://bit.ly/1m4OamO.

Shand, Hope. 2012. "The Big Six: A Profile of Corporate Power in Seeds, Agrochemicals & Biotech." *The Heritage Farm Companion*, Summer, Decorah, IA: Seed Savers Exchange, 10–15.

Shand, Hope, and Pat Mooney. 1998. "Terminator Seeds Threaten an End to Farming." *Earth Island Journal* 13 (4): 30–32.

Sharp, Renee, and Bill Walker. 2007. *Power Drain: Big Ag's X Million Energy Subsidy*. Washington, DC: Environmental Working Group.

Shen, Aviva. 2013. "Congressman Who Gets Millions in Farm Subsidies Denounces Food Stamps as Stealing Other People's Money." *Think Progress*. http://bit.ly/1tU1jxx.

Shepherd, William G., and Joanna M. Shepherd. 2004. *The Economics of Industrial Organization*. 5th ed. Long Grove, IL: Waveland Press.

Shields, Dennis A. 2009. *Dairy Market and Policy Issues*. R40205. Washington, DC: Congressional Research Service.

Shields, Dennis A. 2010. *Consolidation and Concentration in the U.S. Dairy Industry*. R41224. Washington, DC: Congressional Research Service.

Shin, Andy. 2011. *Beer Wars: Public Interest v. Economic Theory*. Greenville, SC: Furman University.

Shlacter, Barry. 2009. "Grocers Irked to Find out Soymilk Nonorganic." *Star-Telegram*, November 8. http://www.star-telegram.com/news/story/1746193.html.

Sieker, Blake. 2009. *Consolidation Direction: Where and Why the Seed Industry Is Headed*. West Des Moines, IA: The Context Network. http://bit.ly/1s78PD7.

Silverstein, Barry. 2011. "Oh Soy: Dairy Fans Urged to Switch from Milk to Silk." *Brandchannel*, March 29. http://bit.ly/1s791Cb.

Skariachan, Dhanya. 2013. "Costco Gains Market Share; Profit Tops Street View." *Reuters*, March 12. http://reut.rs/1pmh7Ze.

Sligh, Michael, and Carolyn Christman. 2003. *Who Owns Organic? The Global Status, Prospects, and Challenges of a Changing Organic Market*. Pittsboro, NC: Rural Advancement Foundation International-USA.

Smith, Andrea. 2009. "The Revolution Will Not Be Funded: Introduction." In *The Revolution Will Not Be Funded: Beyond the Non-Profit Industrial Complex*, edited by INCITE! Cambridge, MA: South End Press.

Smith, Vincent H., Bruce A. Babcock, and Barry K. Goodwin. 2012. *Field of Schemes the Taxpayer and Economic Welfare Costs of Shallow-Loss Farming Programs*. Washington, DC: American Enterprise Institute for Public Policy Research.

Smith, Willie, and Hayden Montgomery. 2004. "Revolution or Evolution? New Zealand Agriculture Since 1984." *GeoJournal* 59 (2): 107–18.

Snyder, Charles E. 2011. *NCB Co-Op Top 100*. Arlington, VA: National Cooperative Bank.

So, Adrienne. 2013. "Your Local Beer isn't as Local as You Think." *Slate*, August 13. http://slate.me/1wuFI3m.

Soil Association. 2013. *Organic Market Report*. Bristol, UK. http://bit.ly/1s795BU.

Souza, Kim. 2014. "Wal-Mart Opens First Walmart to Go Convenience Store." *The City Wire*, March 16. http://www.thecitywire.com/node/32205.

Spaniolo, Lia, and Phil Howard. 2010. "Emerging Eco-Labels: Researching Perspectives of Co-Op Member-Owners." *Cooperative Grocer*, May/June.

Sparks, Janet. 2012. "ABA Forum Presenter Names Top Legal Trends in 2012." *Blue MauMau Franchise News for the Franchisee*, December 25. http://bit.ly/1uD5388.

Stanford, Duane. 2014. "Pabst Blue Ribbon Deal Doesn't Involve Russian Brewer After All." *Bloomberg*, November 18. http://bloom.bg/1Kk5FXv.

Starmer, Elanor, and Timothy A. Wise. 2007. *Living High on the Hog: Factory Farms, Federal Policy, and the Structural Transformation of Swine Production*. Working Paper 07–04. Medford, MA: Global Development and Environment Institute, Tufts University.

Starmer, Elanor, Aimee Witteman, and Timothy A. Wise. 2006. *Feeding the Factory Farm: Implicit Subsidies to the Broiler Chicken Industry*. Working Paper 06–03. Medford, MA: Global Development and Environment Institute, Tufts University.

Starrs, Sean. 2013. "State and Capital: False Dichotomy, Structural Super-Determinism and Moving beyond." In *The Capitalist Mode of Power: Critical Engagements with the Power Theory of Value*, edited by Tim Di Muzio, 117–33. New York: Routledge.

Statista. 2014a. "Revenue of the United States Fast Food Restaurant Industry from 2002 to 2018." http://bit.ly/1qCxwvC.

Statista. 2014b. "McDonald's Corporation Advertising Spending in the United States from 2009 to 2013." *Statista*. http://bit.ly/1uObyEg.

Steltenpohl, Greg. 2005. "Associate Economics." *EnlightenNext Magazine*, May.

Stephens, Arran. 2008. KUOW, Seattle, WARadio. March 5.

Stevens, Sarah, and Peter Jenkins. 2014. *Heavy Costs: Weighing the Value of Neonicotinoid Insecticides in Agriculture*. Washington, DC: Center for Food Safety.

Stobbe, Mike. 2012. "Kids' Cholesterol Down; Fewer Trans Fats Cited." *USA Today*, August 8. http://usat.ly/1y4C9CS.

Striffler, Steve. 2002. "Inside a Poultry Processing Plant: An Ethnographic Portrait." *Labor History* 43 (3): 305–13.

Stromberg, Joseph R. 2001. "The Role of State Monopoly Capital in the American Empire." *Journal of Libertarian Studies* 15 (3): 57–93.

Stuart, Diana. 2011. "'Nature' Is Not Guilty: Foodborne Illness and the Industrial Bagged Salad." *Sociologia Ruralis* 51 (2): 158–74.

Sunlight Foundation. 2014. *Influence Explorer: Cargill Inc*. http://bit.ly/1uObBA3.

Supermarket News. 2007. "Organic Food Sales Grew Twenty-Two Percent in 2006," May 8. http://supermarketnews.com/meals/organic-food-sales-grew-22-2006.

Sussman, Mary. 2014. "The Open-Source Seed Movement in Wisconsin." *Isthmus*, February 20. http://www.thedailypage.com/isthmus/article.php?article=42096.

Swanson, Abbie Fentress. 2014. "Small Farmers aren't Cashing in with Wal-Mart." *National Public Radio*. http://n.pr/19KK8Kb.

Taras, Howard L., and Miriam Gage. 1995. "Advertised Foods on Children's Television." *Archives of Pediatrics & Adolescent Medicine* 149 (6): 649–52.

Taylor, J. Edward, and Philip L. Martin. 1997. "The Immigrant Subsidy in US Agriculture: Farm Employment, Poverty, and Welfare." *Population and Development Review* 23 (4): 855–74.

Taylor, Warren. 2014. "Snowville Creamery's New Community Capitalism." *The Milkweed*, March.

Terlep, Sharon. 2013. "Goldman Left Out of Smithfield Deal." *Wall Street Journal*, June 20. Markets. http://on.wsj.com/1wuG8a8.

Tesoriero, Heather Won, and Peter Lattman. 2006. "How a Tiny Law Firm Made Hay Out of Tainted Spinach." *Wall Street Journal*, September 27. http://on.wsj.com/1k4Zm0p.

The Onion. 2013. "Amazon CEO Buys Washington Post." August 6. http://bit.ly/1uObDIh.

Then, Christoph. 2005. *Monsanto's Pig Monopoly*. Hamburg, Germany: Greenpeace.

Then, Christoph, and Ruth Tippe. 2009. *The Future of Seeds and Food Under the Growing Threat of Patents and Market Concentration*. Hamburg, Germany: No Patents on Seeds Coalition.

Then, Christoph, and Ruth Tippe. 2011. *Seed Monopolists Increasingly Gaining Market Control*. Hamburg, Germany: No Patents on Seeds Coalition.

Thomas, Maria Ajit, and Siddharth Cavale. 2013. "Acquisitive Flowers Seeks Bigger Slice of U.S. Bread Market." *Reuters*, February 21. http://reut.rs/1s2M0pU.

Thompson, Paul B. 2007. "Agriculture and Working-Class Political Culture: A Lesson from The Grapes of Wrath." *Agriculture and Human Values* 24 (2): 165–77.

Tremblay, Victor J., and Carol Horton Tremblay. 2007. "Brewing: Games Firms Play." In *Industry and Firm Studies*, 4th ed., edited by Victor J. Tremblay and Carol Horton Tremblay, 53–79. Armonk, NY: M.E. Sharpe, Inc.

Tsing, Anna. 2009. "Supply Chains and the Human Condition." *Rethinking Marxism* 21 (2): 148–76.

Tuttle, Brad. 2013. "Lawsuit Says Anheuser-Busch Beers Are Even More Watered Down Than You Think." *Time*, February 27. http://ti.me/1m7gi8h.

Tyson Foods. 2014. "Fiscal 2013 Facts." http://bit.ly/1m7gioM.

Urda, Anne. 2006. "Monsanto Sprayed with Herbicide Antitrust Suit." *Law360*, September 28. http://bit.ly/1nWtxob.

USDA. 2013. "National Count of Farmers Market Directory Listing Graph: 1994–2013." *USDA Agricultural Marketing Service*. http://1.usa.gov/1nP0tTM.

USDA. 2014a. "2012 Census Highlights." http://www.agcensus.usda.gov/Publications/2012/Online_Resources/Highlights/.

USDA. 2014b. "Protecting Organic Crops from Synthetic Fertilizers." http://www.ams.usda.gov/AMSv1.0/NOPProtectingOrganicCrops.

USDA. 2014c. "Milk: Production by Year, US." http://nass.usda.gov/Charts_and_Maps/Milk_Production_and_Milk_Cows/milkprod.asp.

U.S. Soybean Export Council. 2011. *How the Global Oilseed and Grain Trade Works*. St. Louis, MO: U.S. Soybean Export Council.

van Bommel, Koen, and André Spicer. 2011. "Hail the Snail: Hegemonic Struggles in the Slow Food Movement." *Organization Studies* 32 (12): 1717–44.

van der Sluis, Wiebe. 2012. "Groupe Grimaud President Frédéric Grimaud: 'A Dream Come True.'" *World Poultry*, May 4. http://bit.ly/1DQwKxc.

Van Munching, Philip. 1997. *Beer Blast: The Inside Story of the Brewing Industry's Bizarre Battle's for Your Money*. New York: Times Books.

Vander Dussen, Sybrand. 2008. "My View on Milk Production Increases." *Milk Producers Council Newsletter*, November 7. http://milkproducerscouncil.org/syps_article.htm.

Vander, Stichele, Myriam. 2012. *Challenges for Regulators: Financial Players in the Food Commodity Derivatives Markets*. Amsterdam: Stichting Onderzoek Multinationale Ondernemingen (SOMO).

Velasco, Schuyler. 2012. "Pink Slime Bankruptcy: After the Backlash, What's Next for Beef?" *Christian Science Monitor*, April 2. http://bit.ly/1I4hzXv.

Velasco, Schuyler. 2013. "McDonald's Helpline to Employee: Go on Food Stamps." *Christian Science Monitor*, October 24. http://bit.ly/KV5UP1.

Vileisis, Ann. 2008. *Kitchen Literacy: How We Lost Knowledge of Where Food Comes from and Why We Need to Get It Back*. Washington, DC: Island Press.

Vitali, Stefania, James B. Glattfelder, and Stefano Battiston. 2011. "The Network of Global Corporate Control." *PLoS ONE* 6 (10): e25995.

Vorley, Bill. 2003. *Food, Inc.: Corporate Concentration from Farm to Consumer*. London: UK Food Group.

Vos, Timothy. 2000. "Visions of the Middle Landscape: Organic Farming and the Politics of Nature." *Agriculture and Human Values* 17 (3): 245–56.

Wagenhofer, Erwin. 2005. *We Feed the World*. http://www.we-feed-the-world.at/en/film.htm.

Walker, Renee E., Christopher R. Keane, and Jessica G. Burke. 2010. "Disparities and Access to Healthy Food in the United States: A Review of Food Deserts Literature." *Health & Place* 16 (5): 876–84.

Walker, Ryan. 2009. *Conserving Turkeys of Thanksgivings Past for the Future*. Pittsboro, NC: The Livestock Conservancy.

Wallace, Alicia. 2013. "Q&A with Steve Demos, Founder of WhiteWave." *Boulder Daily Camera*, May 23. http://bit.ly/1sUzMhA.

Wallack, Lawrence, Diana Cassady, and Joel Grube. 1990. *TV Beer Commercials and Children: Exposure, Attention, Beliefs, and Expectations About Drinking as an Adult.* Washington, DC: AAA Foundation for Traffic Safety.

Waltz, Emily. 2010. "Monsanto Relaxes Restrictions on Sharing Seeds for Research." *Nature Biotechnology* 28 (10): 996.

Wang, Leyi, Beverly Byrum, and Yan Zhang. 2014. "New Variant of Porcine Epidemic Diarrhea Virus, United States, 2014." *Emerging Infectious Diseases* 20 (5): 917–19.

Ward, Clement E. 2010. "Assessing Competition in the U.S. Beef Packing Industry." *Choices* 25 (2): 1–14.

Warner, Melanie. 2013. *Pandora's Lunchbox: How Processed Food Took Over the American Meal.* New York: Simon and Schuster.

Waterfield, Larry. 2009. "European Union Farm Subsidies Come with a Cost." *The Packer*, August 28. http://bit.ly/1m7gogd.

Waters, Christina. 1996. "Seeds of the Future." *Metro*, September 5. http://www.metroactive.com/papers/metro/09.05.96/produce-9636.html.

Watson, Elaine. 2014. "Hain Celestial Buys Rudi's Organic Bakery in $61.3 Million Deal." *FoodNavigator-USA.com*, April 28. http://bit.ly/1n6Y0AZ.

Watts, Michael J. 1994. "Epilogue: Contracting, Social Labor, and Agrarian Transitions." In *Living Under Contract: Contract Farming and Agrarian Transformation in Sub-Saharan Africa*, edited by Peter D. Little and Michael J. Watts, 248–57. Madison, WI: University of Wisconsin Press.

Weinraub, Judith. 2003. "Who Pulled a Fast One on the Organic Food Industry?" *Washington Post*, March 19, F1.

Weis, Tony. 2010. "The Accelerating Biophysical Contradictions of Industrial Capitalist Agriculture." *Journal of Agrarian Change* 10 (3): 315–41.

Welsh, Rick, Chantal Line Carpentier, and Bryan Hubbell. 2001. "On the Effectiveness of State Anti-Corporate Farming Laws in the United States." *Food Policy* 26 (5): 543–48.

Welsh, Rick, Bryan Hubbell, and Chantal Line Carpentier. 2003. "Agro-Food System Restructuring and the Geographic Concentration of US Swine Production." *Environment and Planning A* 35 (2): 215–29.

Whole Foods. 2014. "Unacceptable Ingredients for Food." *Whole Foods Market*. http://bit.ly/1bn2OzT.

Wilde, Matthew. 2009. "Independent Seed Companies a Dying Breed." *Waterloo Cedar Falls Courier*, May 31. http://bit.ly/1m7guo8.

Wilde, Parke. 2013a. *Food Policy in the United States: An Introduction.* New York: Routledge.

Wilde, Parke. 2013b. "U.S. Food Policy: Long-Hidden Details Revealed about the Pork Checkoff Program's Sixty Million Dollar Purchase." *U.S. Food Policy*. http://usfoodpolicy.blogspot.com/2013/02/long-hidden-details-revealed-about-pork.html.

Wilkinson, John. 2009. "Globalization of Agribusiness and Developing World Food Systems." *Monthly Review* 61 (4): 38–49.

Williams, J., P. Scarborough, A. Matthews, G. Cowburn, C. Foster, N. Roberts, and M. Rayner. 2014. "A Systematic Review of the Influence of the Retail Food Environment Around Schools on Obesity-Related Outcomes." *Obesity Reviews* 15 (5): 359–74.

Williams, Sean. 2012. "Tales from the Crypt: More Horrifying CEO Pay Packages." *The Motley Fool*. http://bit.ly/1ynkLnF.

Wilson, Jacque. 2013. "Kraft Removing Artificial Dyes from Some Mac and Cheese." *CNN*, November 4. http://cnn.it/1m7gwfJ.

Winders, Bill. 2009. *The Politics of Food Supply: U.S. Agricultural Policy in the World Economy*. New Haven, CT: Yale University Press.

Wingfield, Nick, and Ben Worthen. 2010. "Copycat Farmers' Markets Reap a Crop of Complaints." *Wall Street Journal*, September 24. http://on.wsj.com/1AONAdQ.

Winson, Anthony. 2013. The Industrial Diet: The Degradation of Food and the Struggle for Healthy Eating. Vancouver, BC, Canada: UBC Press.

Winters, Patrick. 2012. "Syngenta to Buy Biotech Seedmaker Devgen for $523 Million." *Bloomberg*, September 21. http://bloom.bg/X4Af3P.

Wisdorff, Armin. 2013. "EU Farm Policy 2014-2020: MEPs Give Final Blessing to Greener and Fairer CAP." *European Parliament News*, November 20. http://bit.ly/1pg6Zy3.

Wise, Timothy A. 2004. *The Paradox of Agricultural Subsidies: Measurement Issues, Agricultural Dumping, and Policy Reform*. Working Paper 04–02. Medford, MA: Global Development and Environment Institute, Tufts University.

Wise, Timothy A. 2013. *Can We Feed the World in 2050? A Scoping Paper to Assess the Evidence*. Working Paper 13–04. Medford, MA: Global Development and Environment Institute, Tufts University.

Wise, Timothy A., and Sarah E. Trist. 2010. *Buyer Power in U.S. Hog Markets: A Critical Review of the Literature*. Working Paper 10–04. Medford, MA: Global Development and Environment Institute, Tufts University.

Wohl, Jessica, and Ellen Jean Hirst. 2014. "Tyson, Hillshire Reach Deal with Justice Dept. on Takeover." *Chicago Tribune*, August 27, http://trib.in/1y4D3zl.

Wood, Ellen Meiksins. 1998. "The Agrarian Origins of Capitalism." *Monthly Review* 50 (3): 14–31.

Wood, Steve. 2013. "Revisiting the US Food Retail Consolidation Wave: Regulation, Market Power and Spatial Outcomes." *Journal of Economic Geography*, 13 (2): 203–210.

Wrigley, Neil. 2001. "The Consolidation Wave in U.S. Food Retailing: A European Perspective." *Agribusiness* 17 (4): 489–513.

Wrigley, Neil. 2002. "Transforming the Corporate Landscape of US Food Retailing: Market Power, Financial Re-Engineering and Regulation." *Tijdschrift Voor Economische En Sociale Geografie* 93 (1): 62–82.

Wrong, Dennis Hume. 1995. *Power: Its Forms, Bases, and Uses*. New Brunswick, NJ: Transaction Publishers.

Wulfraat, Marc. 2014. *Direct Store Delivery versus Centralized Distribution*. Montreal, Canada: MWPVL. http://www.mwpvl.com/html/dsd__vs_central_distribution.html.

Yamaguchi, Yuki. 2013. "7-Eleven Owner to More Than Double North America Stores." *Business Week*, June 4. http://buswk.co/1tV3RiC.

Ylisela, Jr., James, David Sterrett, and Kate MacArthur. 2010. "Pay-to-Play Infects Chicago Beer Market, Crain's Investigation Finds." *Crain's Chicago Business*, November 22.

York, Emily Bryson. 2010. "Sara Lee Latest to Curtail Use of High-Fructose Corn Syrup." *Chicago Tribune*, August 17. http://trib.in/1ks3ZAR.

Zachary, G. Pascal. 1999. "Many Industries Are Congealing into Lineup of Few Dominant Giants." *Wall Street Journal*, March 8, B1.

Index